U.S. Department of Transportation
National Highway Traffic Safety Administration

DOT HS 810 863
DOT-VNTSC-NHTSA-07-02

November 2007

A Safety Roadmap for Future Plastics and Composites Intensive Vehicles

This document is available to the public from the National Technical Information Service, Springfield, Virginia 22161

Notice

This document is disseminated under the sponsorship of the Department of Transportation in the interest of information exchange. The United States Government assumes no liability for its contents or use thereof.

REPORT DOCUMENTATION PAGE			*Form Approved* *OMB No. 0704-0188*

Public reporting burden for this collection of information is estimated to average 1 hour per response, including the time for reviewing instructions, searching existing data sources, gathering and maintaining the data needed, and completing and reviewing the collection of information. Send comments regarding this burden estimate or any other aspect of this collection of information, including suggestions for reducing this burden, to Washington Headquarters Services, Directorate for Information Operations and Reports, 1215 Jefferson Davis Highway, Suite 1204, Arlington, VA 22202-4302, and to the Office of Management and Budget, Paperwork Reduction Project (0704-0188), Washington, DC 20503.

1. AGENCY USE ONLY (Leave blank)	2. REPORT DATE November 2007	3. REPORT TYPE AND DATES COVERED Final Report November 2007	
4. TITLE AND SUBTITLE A Safety Roadmap for Future Plastics and Composites Intensive Vehicles			5. FUNDING NUMBERS HS53/DG066
6. AUTHOR(S) Aviva Brecher, Ph.D., National Technical Expert, Policy and Planning Division			
7. PERFORMING ORGANIZATION NAME(S) AND ADDRESS(ES) U.S. Department of Transportation Research and Innovative Technology Administration John A. Volpe National Transportation Systems Center Advanced Safety Technology Division 55 Broadway Cambridge, MA 02142			8. PERFORMING ORGANIZATION REPORT NUMBER DOT-VNTSC-NHTSA-07-02
9. SPONSORING/MONITORING AGENCY NAME(S) AND ADDRESS(ES) U.S. Department of Transportation National Highway Traffic Safety Administration 1200 New Jersey Avenue SE Washington, DC 20590			10. SPONSORING/MONITORING AGENCY REPORT NUMBER DOT HS 810 863
11. SUPPLEMENTARY NOTES Volpe National Transportation Systems Center, RITA, U.S. DOT			
12a. DISTRIBUTION/AVAILABILITY STATEMENT This document is available free of charge from the NHTSA Web site at: http://www.nhtsa.dot.gov/staticfiles/DOT/NHTSA/NRD/Multimedia/Crashworthiness/4680PCIV_SafetyRoadmap-Nov2007.pdf			12b. DISTRIBUTION CODE
13. ABSTRACT (Maximum 200 words) This report summarizes the approach, activities, and results of a study to evaluate the potential safety benefits of Plastics and Composites Intensive Vehicles (PCIVs), to enable their deployment by 2020. The main goals were to review and assess the state of knowledge in order to identify gaps, key research needs, and the challenges and opportunities for safety enhancements. PCIV-related safety enhancements that could benefit an aging driver population were selected as a priority research focus. The Situation Analysis was conducted, based on a review of technical literature, national research efforts on automotive light-weighting and the crash safety performance of advanced materials, complemented by a focused survey of diverse subject matter experts. The analysis identified near-term, midterm, and long-term research needs and priorities to facilitate future PCIV deployment. A PCIV Safety Roadmap was developed, which synthesizes the study findings and outlines follow-on research and milestones to measure progress towards the design, development and technology integration of fuel-efficient and safe PCIVs by 2020.			
14. SUBJECT TERMS Automotive crash safety; plastics and composites intensive vehicles (PCIVs); light-weighting advanced materials; older driver safety; technology integration roadmap; public private partnership; crash safety standards; integrated safety strategy			15. NUMBER OF PAGES 104
			16. PRICE CODE
17. SECURITY CLASSIFICATION OF REPORT Unclassified	18. SECURITY CLASSIFICATION OF THIS PAGE Unclassified	19. SECURITY CLASSIFICATION OF ABSTRACT Unclassified	20. LIMITATION OF ABSTRACT Unlimited

NSN 7540-01-280-5500

PREFACE AND ACKNOWLEDGEMENTS

The Volpe National Transportation Systems Center (Volpe Center) of the U.S. Department of Transportation's (USDOT) Research and Innovative Technology Administration (RITA) provides research, evaluation, and technical analysis support to the National Highway Traffic Safety Administration's (NHTSA) Office of Vehicle Safety Research to assess, develop, and implement research strategies for improving vehicle safety.

In November 2005, the American Plastics Council, in cooperation with NHTSA, sponsored a Technology Integration Workshop on "Enhancing Future Automotive Safety With Plastics," and published its findings in a Technology Integration Report in May 2006. The major opportunities and challenges for enhancing the safety of next generation vehicles using advanced plastics and composite materials in structural and safety applications were identified, as well as the need to develop a safety roadmap. The FY06 Transportation Appropriations Senate Report 109-109 provided both the resources and the guidance to NHTSA to explore the potential safety benefits of light-weight, fuel efficient Plastics and Composites Intensive Vehicles (PCIVs), and develop the foundation for research cooperation with the Department of Energy, industry, universities and other safety stakeholders. NHTSA tasked the Volpe Center to assess the current state of knowledge and emerging safety technology opportunities to enhance the crash safety of PCIVs by 2020, with special emphasis on improved protection of older drivers and occupants.

This report summarizes the results of a foundational study undertaken in support of NHTSA to identify the research needs and priorities in the near-, mid-, and long-term, as well as the challenges and opportunities to the development and commercial deployment of light-weight, plastics-rich, and fuel efficient vehicles by 2020 that would be designed to comply with all applicable crash safety standards and regulatory requirements. Taking an international and national perspective, the Volpe Center reviewed relevant literature, interviewed subject matter experts and ongoing research, developed a PCIV 2020 safety Vision, and identified the major R&D needs, challenges, and opportunities to develop a roadmap for PCIV safety.

The primary author of the report is Dr. Aviva Brecher, Principal Investigator, with technical support from Dr. John Brewer, as the expert on materials science, composites testing standards, and vehicle crash safety. Other Volpe Center managers and colleagues who contributed to many aspects of this study and provided technical review and quality assurance include: Kevin Green, Chief of Advanced Safety Technology Division; Dr. Wassim Najm for issues related to NHTSA integrated safety technologies and strategy; Samuel Toma and Emily Lumley for assistance with graphics, report formatting, and organization; and John O'Donnell, Director, and Robert Dorer, Deputy Director, Office of Surface Transportation, for their continued support and guidance.

Special appreciation is due to Dr. William Thomas Hollowell, NHTSA Director of Vehicle Safety Research, and Sanjay Patel, NHTSA project manager, for their effective oversight and timely review. Dr. David Smith, Chief, Structures and Restraints Research Division, and Barbara Hennessey, manager of the Hydrogen Vehicle Safety Research Program are thanked for providing technical support, feedback, and guidance as needed. Thanks are due to all NHTSA regulatory, policy, and technical staff who conducted an agency-wide review and provided helpful comments to improve this report.

The American Chemistry Council- Plastics Division (ACC-PD) industry partners supported this safety-oriented study with insights from previous technology integration roadmaps, and provided technical review comments and inputs throughout the process. They are Dr. Michael Fisher, Technical Director, Mr. James Kolb, Communications Director of the Automotive Center, and Ms. Suzanne Cole, President of Cole and Associates, Inc. The Volpe team made several presentations to NHTSA and ACC-PD collaborators and engaged them in regular telephone conferences to obtain feedback.

Valuable inputs are gratefully acknowledged from the Department of Energy (DOE) managers of the FreedomCAR partnership, the DOE National Laboratories Research program managers, the industry principals involved in cooperative USCAR R&D efforts, and Subject Matter Experts (SMEs) from professional, standards, and trade associations, industry, and academia. Those who agreed to be interviewed and generously shared their insights and publications are listed by name in Appendix 4.2 and cited or mentioned in footnotes.

LIST OF ACRONYMS

AARP	American Association For Retired Persons
ABS	Antilock Brake Systems
ACC	Automotive Composites Consortium, affiliated with USCAR
ACC-PD	American Chemistry Council- Plastics Division
ACE	Advanced Compatibility Engineering
AFVs	Alternative Fueled Vehicles
APC	American Plastics Council (renamed ACC-PD)
ASTM	American Society for Testing and Materials
ATP	Advanced Technology Program
AXP	Automotive X-Prize
BIW	Body In White
BSP	Best Safety Practices
CAFE	Corporate Average Fuel Economy
CCC	Chrysler Composite Car
CEM	Crash Energy Management
CEMWG	Crash Energy Management Working Group
CFRC	Carbon Fiber Reinforced Composites
CIREN	Crash Injury Research And Engineering Network
CoE	Centers of Excellence
DARPA	Defense Advanced Research Program Agency
DOC	Department of Commerce
DOE	Department of Energy
EMC	Electromagnetic Compatibility
EOL	End-of-Life
ESC	Electronic Stability Control
EU	European Union
FAA	Federal Aviation Administration
FCVs	Hydrogen Fuel Cell Vehicles
FCVT	FreedomCar And Vehicle Technologies
FFVs	Flexi-Fueled Vehicles
FHWA	Federal Highway Administration
FMVSS	Federal Motor Vehicle Safety Standards
HPS	High Performance Steel
HPSS	High Performance Stainless Steels
HUD	Heads Up Displays
IIHS	Insurance Institute for Highway Safety
ISO	International Organization for Standardization
ITS	Intelligent Transportation Systems
IVBSS	Integrated Vehicle Based Safety Systems
MEP	Manufacturing Extension Program
METI	Ministry of Economy, Trade, and Industry
VMT	Vehicle Miles Traveled

LIST OF ACRONYMS

NHTSA	National Highway Traffic Safety Administration
NIST	National Institute Of Standards And Technology
NRC	National Research Council
NRL	Naval Research Lab
OEMs	Original Equipment Manufacturers
OSRP	Occupant Safety Research Partnership
PCIV	Plastics And Composite Intensive Vehicles
P3	Public-Private Partnership
PMC	Polymer Matrix Composites
PNGV	Partnership For A New Generation Of Vehicles
PNNL and ORNL	Department Of Energy Labs
R&D	Research And Development
RD&T	Research, Development, And Technology
RDT&E	Research, Development, Test, And Evaluation
RFI	Radio Frequency Interference
RITA	Research And Innovative Technology Administration
SAE	Society Of Automotive Engineers
SEAS	Secondary Energy Absorbing Structures
SPE	Society Of Plastics Engineers
T&E	Test And Evaluation
TARDEC	Tank Automotive Research, Development, And Engineering Center
TECABS	Technologies For Carbon Fiber Reinforced Modular Automotive Body Structures
TVMT	Truck Vehicle Miles Traveled
ULSAB	Ultra Light Steel Auto-body
USAMP	US Automotive Materials Partnership
USCAR	United States Council For Automotive Research
USDOT	U.S. Department Of Transportation
UTC	University Transportation Centers
VOC	Volatile Organic Compounds
VRHS	Variable Ride-Height Suspension
VRP	Vehicle Recycling Partnership
WHIPS	Whiplash Protection System

TABLE OF CONTENTS

1. **BACKGROUND AND INTRODUCTION** 1
   1.1 THE 2006 NHTSA PCIV RESEARCH INITIATIVE AND NATIONAL SUSTAINABILITY GOALS 1
   1.2 THE AMERICAN PLASTICS COUNCIL WORKSHOP: RECOMMENDATIONS TO ENHANCE PCIV SAFETY 5
   1.3 PROJECT SCOPE AND TECHNICAL APPROACH 6

2. **A 2020 VISION FOR PCIV SAFETY** 8
   2.1 RESOURCES FOR A 2020 PCIV SAFETY VISION STATEMENT 8
      2.1.1 The Department of Transportation (DOT) Strategic Plan, 2006-2011 8
      2.1.2 The DOT Strategic Research Plan 8
      2.1.3 NHTSA Vehicle Safety Research Plans and Programs 9
      2.1.4 The NHTSA Integrated Vehicle Safety Research Strategy 10
   2.2 THE NEED TO ENHANCE THE SAFETY OF FUTURE VEHICLES AND OLDER DRIVER PROTECTION 12
      2.2.1 Crash protection needs of older drivers and occupants 12
      2.2.2 The need to capture the potential safety benefits of PCIVs 17
   2.3 VISION STATEMENT FOR PCIV SAFETY 19
      2.3.1 Resources for PCIV Safety Vision 19
      2.3.2 PCIV Safety Research Goals and Objectives 20
      2.3.3 Performance Metrics and Milestones 20

3. **SITUATION ANALYSIS-APPROACH AND FINDINGS** 22
   3.1 TECHNICAL APPROACH 22
   3.2 THE ACC-PD SAFETY PRIORITIES FOR FUTURE PCIVs 23
   3.3 EXISTING RESEARCH PARTNERSHIPS RELEVANT TO PCIV SAFETY 25
      3.3.1 DOT research relevant to automotive composites and safety performance 25
         3.3.1.1 NHTSA Research Programs 25
         3.3.1.2 FAA Research Programs 26
         3.3.1.3 FHWA Research Programs 27
         3.3.1.4 FTA Research Programs 27
         3.3.1.5 The DOT University Transportation Centers 27
         3.3.1.6 Small Business Innovation Research (SBIR) Program 27
      3.3.2 Public-Private Partnership (P3) Research and Development Programs 27
      3.3.3 Other Federal R&D Related to PCIV 30
   3.4 STANDARDS AND GUIDELINES FOR AUTOMOTIVE COMPOSITES CRASHWORTHINESS 31
      3.4.1 The Society of Automotive Engineers International (SAE) 31
      3.4.2 The American Society for Testing and Materials (ASTM) 32
      3.4.3 ISO TC 61/SC 13 33
   3.5 INTERNATIONAL RESEARCH AND DEVELOPMENT EFFORTS ON AUTOMOTIVE LIGHT-WEIGHTING WITH COMPOSITES 33
      3.5.1 European Union Research and Development Partnerships 33

      3.5.2 Research and Development Partnerships in Japan 34
  3.6 CURRENT TRENDS, BEST PRACTICES AND LESSONS LEARNED FOR AUTOMOTIVE COMPOSITES INTEGRATION.. 35
      3.6.1 Emerging PCIV Concepts and Best Safety Practices (BSP) 35
      3.6.2 Lessons Learned 38

4. SURVEY OF EXPERTS AND SUMMARY OF FINDINGS 41
  4.1 THE EXPERTS' SURVEY DESIGN 41
  4.2 SUMMARY OF EXPERTS' INPUTS ON PCIV SAFETY PRIORITIES 42
      4.2.1 Knowledge Gaps in Predicting the Crash Performance of Plastics and Composites: 42
      4.2.2 Research Needs to Predict the Crashworthiness of Automotive Composites 44
      4.2.3 The American Chemistry Council-Plastics Division Survey 44
      4.2.4 Priority Research Opportunities for Future PCIV Safety 45
      4.2.5 Research Needs For Occupant Safety 46
      4.2.6 Barriers and challenges to PCIV development and deployment 47
      4.2.7 Suggested Strategies to Overcome Barriers to PCIV Safety Deployment 48
      4.2.8 Suggested NHTSA Role and Opportunities for PCIV Safety R&D 48
  4.3 RECOMMENDED TOP-3 PCIV SAFETY RESEARCH AND DEVELOPMENT AND TECHNOLOGY INTEGRATION PRIORITIES FOR ROADMAP DEVELOPMENT 49
      4.3.1 Near-term Priorities (3-5 years) for Research, Development and Technology 49
      4.3.2 Mid-term priorities (5-10 years) for Test and Evaluation 50
      4.3.3 Long-term priorities (10-15 years) 50

5. SAFETY ROADMAP FOR FUTURE PCIVS 51
  5.1 THE 2020 PCIV SAFETY ROADMAP DEVELOPMENT STRATEGY 51
  5.2 BUILDING ON EXISTING ROADMAPS FOR PCIV SAFETY ROADMAP 52
      5.2.1 The DOE/USCAR and FreedomCAR Roadmaps 52
      5.2.2 The NHTSA Integrated Safety Strategy and Timeline 61
      5.2.3 Extending the ACC-PD Automotive Technology Roadmaps to PCIV Safety 61
  5.3 TRANSLATING R&D PRIORITIES INTO PCIV SAFETY ROADMAPS 63
  5.4 POTENTIAL NHTSA ROLE IN SAFETY ASSURANCE OF FUTURE PCIVS 63

6. APPENDICES 70
  6.1 APPENDIX 3.1- PRIORITY CROSSCUT SAFETY ISSUES IN THE APC WORKSHOP REPORT ENHANCING AUTOMOTIVE SAFETY WITH PLASTICS 70
  6.2 APPENDIX 4.1- EXPERTS INTERVIEW GUIDE 73
  6.3 APPENDIX 4.2-LIST OF EXPERTS INTERVIEWED 75

7. REFERENCES 76

LIST OF FIGURES

Figure 1-1: Comparative data on materials substitution options for light-weighting vehicles to improve fuel efficiency (from Carpenter, 2006). 5

Figure 2-1: NHTSA Strategic R&D programs support its safety regulatory mission. 11

Figure 2-2: NHTSA Vision Statement (W.T. Hollowell "Overview of NHTSA Research for Enhancing Safety," 11.10. 2006, Presentation to MADYMO International Users Meeting in Detroit, MI.) 11

Figure 2-3: From the U.S. Bureau of Census special report "65+ in the United States: 2005"..................12

Figure 2-4: Total fatalities and normalized rates for the over 70 drivers versus time. 14

Figure 2-5: Projected totals of aging "baby boomers" by age group, to 2050. 15

Figure 2-6: MIT Age Lab forecasts a 35% increase in VMT for aging drivers by 2020. 16

Figure 2-7: From the Dec. 19, 2006 IIHS Status Report newsletter, "Bigger is generally better" shows that the risk of crash fatality is higher for lighter, more fuel-efficient vehicles at present. 18

Figure 3-1 : DOE Summary of advantages for Carbon Fiber Polymer Composites in automotive and aerospace structural applications and their relatively high crash Energy Absorption (EA), when compared with alternative light weighting materials (see DOE/ USCAR references). 31

Figure 5-1: Timeline, performers and roles for the DOE FreedomCAR Automotive Composites Consortium partnership for materials technology development. 55

Figure 5-2 : Lightweight and high strength materials for applications to structures and propulsion in next generation vehicles are an explicit objective under the FreedomCAR and Fuel Partnerships and the 21st Century Truck Partnership. 56

Figure 5-3: Depiction of the USCAR strategy to model, develop, integrate, verify, and validate emerging lightweight, high-strength materials and technologies for a future reference fuel-efficient vehicle. 57

Figure 5-4: Figure from "Driving Technology: A Transition Strategy to Enhance Energy Security," DOE/EERE May 26, 2006. Safety is an implicit "utility" in every step on the ladder depicting the progress towards energy efficient FCVs over time. 58

Figure 5-4a: Selected Roadmaps for automotive composite materials from "FreedomCAR and FreedomFUEL Partnership: Materials Technology Roadmap," Oct 2006. 59

Figure 5-4b: The structural strength of glazing plastics and composite materials for vehicle windows and roofs are very important to occupant safety in crashes (e.g., to prevent ejection in rollovers). 60

Figure 5-5: Summary of research and development priorities for plastics and composites in automotive safety applications, as identified in the May 2006 ACC-PD report. 64

Figure 5-6: A and B- NHTSA technology integration timeline for active and passive safety technologies.. 65

Figure 5-7 : Roadmap for Predictive Engineering tools needed to model crash safety performance of plastics-intensive vehicles. This is a key mid-term (5-10 years) research and technology priority identified in this study. 66

Figure 5-8: Timeline for Research and Technology Integration for PCIV Safety Roadmap 67

Figure 5-9: Roadmap to PCIV Safety 68

Figure 5-10: Strategic Priority Activities Leading to PCIV Safety Assurance by 2020 69

INDEX OF TABLES

Table 3.1: ACC-PD Priority Technology Integration Activities 24

EXECUTIVE SUMMARY

The Transportation Appropriations Senate Report 109-109 issued guidance to NHTSA to use $250,000 of FY2006 money to "begin development of a program to examine the possible safety benefits of lightweight plastic and composite intensive vehicle (PCIV)," in cooperation with the Department of Energy (DOE) and industry and other automotive safety stakeholders.

NHTSA requested that the Volpe Center conduct a research study in FY2007 to assess the state of knowledge and emerging safety technology opportunities for assuring the crash safety of future lightweight vehicles, including PCIVs, with special emphasis on the improved protection of older drivers and occupants. The project built on the recommendations of a *Technology Integration Report and Prospectus* (May 2006) summarizing the workshop on *Enhancing Future Automotive Safety with Plastics* (Nov. 2005), which was sponsored by the American Plastics Council (APC)[1] in partnership with NHTSA.

The major goals addressed by this study were to:
- Develop a vision statement for 2020 PCIV Safety, with special focus on enhanced crash safety of older drivers, and identify performance goals, metrics, and milestones. This vision is consistent with national priorities for vehicle fuel efficiency, environmental preservation, and safety. It is also synergistic with the NHTSA integrated vehicle safety strategy and its vehicle safety research program (Chapter 2).
- Conduct a structured situation analysis of the automotive plastics and composites in safety applications to identify knowledge gaps, high-priority research needs, barriers to PCIV deployment, and strategies to overcome them. The analysis was based on:
  - The review and evaluation of safety-related components of national and global automotive light-weighting research, as well as technology development and integration efforts (Chapter 3); and
  - Conduct, analyze, and summarize focused interviews of leading experts on crash safety and automotive materials and designs, representing diverse automotive safety stakeholders (Federal agencies and laboratories, industry, academia, technical standards organizations, and nonprofits) (Chapter 4);
- Develop a PCIV safety research and technology roadmap to 2020, which highlights near-term (1-3 years), mid-term (3-5 years) and long-term (to 2020) R&D priorities and milestones. The safety roadmap complements and extends the *Plastics in Automotive Markets: Vision and Technology Roadmap* (APC, 2002) to enable PCIV safety design verification by 2020 (Chapter 5).

This study identified the major opportunities and challenges for enhancing the safety of next generation vehicles using advanced plastics and composite materials, and developed a PCIV Safety Research Roadmap, in order to facilitate the conceptual design, development, and deployment of lightweight, fuel-efficient and environmentally sustainable PCIV which meet or exceed NHTSA crash-safety standards.

---

[1] APC became the American Chemistry Council - Plastics Division (ACC-PD) in December 2006.

1. BACKGROUND AND INTRODUCTION

1.1 The 2006 NHTSA PCIV Research Initiative and National Sustainability Goals

This section provides the background and rationale for this study, undertaken on behalf of the National Highway Safety Administration (NHTSA) in response to Congressional guidance. The FY06 Transportation Appropriations Senate Report 109-109 included the requirement for NHTSA to initiate a foundational cooperative research program on the potential safety benefits of using plastics and composites in the emerging lighter weight, more fuel efficient vehicles:

"**Plastic and Composite Vehicles**—The Committee recognizes the development of plastics and polymer-based composites in the automotive industry and the important role these technologies play in improving and enabling automobile performance. The Committee recommends $250,000 to begin development of a program to examine possible safety benefits of lightweight Plastics and Composites Intensive Vehicles [PCIVs]. The program will help facilitate a foundation between DOT, the Department of Energy and industry stakeholders for the development of safety-centered approaches for future light-weight automotive design."

Congress did not define the plastics content of future PCIVs, but a substantially higher content than at present is necessary to reduce vehicle mass so as to double fuel efficiency.[2]

On January 23, 2007, in the State of the Union address[3], the President launched a new energy "20 in 10" initiative, to reduce domestic gasoline consumption by 20 percent over the next 10 years, through a combination of higher vehicle fuel efficiency and greater use of renewable fuels like ethanol. The new initiative specifies that CAFE standards for passenger vehicles increase 4 percent annually to achieve the 40 mpg fuel efficiency goal by 2017. NHTSA Administrator Nicole Nason addressed the CAFE policy and implementation challenges in her testimony.[4] Proposed pending legislation will enable NHTSA to develop the regulatory framework for improving the fleet-wide fuel efficiency in a manner which is consistent with its public safety goals.

---

[2] The FreedomCAR goal is to "*Dramatically reduce oil consumption by improving the efficiency of personal vehicles and double fuel economy in commercial vehicles.*" See "*Driving Technology: A Transition Strategy to Enhance Energy*" at www.eere.energy.gov/vehiclesandfuels/pdfs/program/tsp_paper_final.pdf - 2006-08-01 - Text Version

[3] See details posted at http://www.whitehouse.gov/stateoftheunion/2007/initiatives/energy.html

[4] Testimony of NHTSA Administrator on March 6, 2007, on CAFÉ standards to the Senate Commerce, Science, and Transportation Committee is posted at http://testimony.ost.dot.gov/test/nason2.htm

In response to the global and national pressures to improve energy efficiency and preserve the environment, Federal and industry initiatives have proposed and adopted multiple strategies and innovative technologies towards a more sustainable, energy efficient, and environmentally cleaner transportation system.

Current research and development (R&D) and test and evaluation (T&E) jointly performed by Federal agencies and industry focus on developing the technology base for more fuel-efficient and environmentally sustainable transportation options, including electric and hybrid propulsion, hydrogen fuel cell vehicles (FCVs), and alternative fuel and flexi-fuel vehicles (AFVs and FFVs). Both the energy efficiency and the safety of future vehicles might benefit from the PCIV research on light-weighting with strong plastics and composites.

To achieve lightweight architectures, without compromising on rigidity, automakers have been researching the replacement of steel with plastics, composites, foams, aluminum, and magnesium. Leading experts[5] have argued that the use of advanced materials for reducing weight offers the easiest and least expensive way to achieve multiple benefits (reduce energy consumption and emissions at equal or better safety). Weight reduction also offers a potentially cost-effective means to reduce fuel consumption and greenhouse gases from the transportation sector. It has been estimated that:
- For every 10 percent reduction in the weight of the total vehicle, fuel economy would improve by 5-7 percent; and
- For every kilogram of vehicle weight reduction, the potential reduction in carbon dioxide emissions is about 20 kg.

> *Vehicle programs designed to achieve major fuel economy improvements must incorporate significant weight savings. The widespread application of lightweight materials and innovative manufacturing processes is necessary to attain this goal. FreedomCAR has set a vehicle weight reduction target of 50 percent, with the additional criterion "affordable cost."*
> Executive Summary, MATERIALS Conclusion from the
> *Review of the Research Program of the FreedomCAR and Fuel Partnership,* First report (NRC, 2005)

The most recent American Chemistry Council-Plastics Division (ACC-PD) data indicate that:
- The average vehicle uses about 150 kg of plastics and composites, versus 1,163 kg (2,559 lbs) of iron and steel[6];
- The automotive industry uses engineered polymer composites and plastics in a wide range of applications, as the second most common class of automotive

---

[5] See Chapter 6 for references by Lovins et al. and Jackson and Schlesinger, 2004.

[6] APC data as cited in "*Plastic on the Outside,*" SAE Automotive Engineering Journal, Aug. 2006, pp. 46-49, at www.aei-online.org and in "Automotive Composites-a Design and Manufacturing Guide," 2nd edition, 2006.

BACKGROUND AND INTRODUCTION

materials after ferrous metals and alloys (cast iron, steel, nickel) which represent 68 percent by weight; other non-ferrous metals used include copper, zinc, aluminum, magnesium, titanium, and their alloys.
- The plastics contents of commercial vehicles comprise about 50 percent of all interior components, including safety subsystems, and door and seat assemblies.
- Industry trends project a substantial increase in use of automotive plastics over the next two decades for reducing vehicle net weight, and for improving environmental impacts and fuel efficiency in response to consumer pressures, and to take advantage of the rapid advances in materials science and technology.

Other options for light-weighting cars for improved fuel efficiency and structural strength include: high performance steel (HPS), which today comprises 26 percent of the average car body, and aluminum, which has half the weight of iron and better corrosion resistance but also has higher costing, energy intensifying, and more difficult manufacturability.

The next generation of fuel-efficient, light-weight vehicles will have to demonstrate full compliance with NHTSA safety regulations for vehicle crashworthiness. Statistical crash data indicate that lighter vehicles have some safety challenges in crashes with heavier ones (a vehicle-to-vehicle compatibility challenge). PCIV designs, materials, and technologies will have to compensate through safety enhancements for equal or better crash performance, in order to gain both public acceptance and market share. At the same time, if smaller, fuel efficient PCIV vehicles are to gain market share in the U.S., the crash compatibility and aggressivity issues of larger vehicles will have to be addressed with advanced design, materials, and technology solutions.

Improved road safety is a major global safety and health concern – the World Health Organization (part of the United Nations) has increased its focus on automotive safety as a growing concern.[7] The ACC Technology Integration Report reflects the plastic industry's strong commitment to enhanced automotive safety and is aligned with both domestic and global needs.

In recognition of the need for safety assurance of the next generation of lightweight, fuel efficient vehicles, the Department of Energy vision statement for the FreedomCAR and Vehicle Technologies (FCVT) program[8] states that "Transportation energy security will be achieved through a U.S. highway vehicle fleet of affordable, full-function cars and trucks that are free from petroleum dependence and harmful emissions **without sacrificing mobility, safety, and vehicle choice.**"

---

[7] World Health Organization (WHO): "*World Report on Road Traffic Injury Prevention*" (2004); Fact sheets on "*Road Safety: a Public Health Issue*"; and "*Road Traffic Injury Prevention Training Manual*" (2006) posted at www.who.int/violence_injury_prevention/publications/road_traffic/en

[8] Details on these research and technology (R&T) partnerships with industry and academia are provided at http://www1.eere.energy.gov/vehiclesandfuels/

Recent reviews of the goals and accomplishments of the FreedomCAR[9] and FUEL initiatives, and of related light-weighting materials research indicate that a major cooperative research effort is devoted to characterizing the Energy Absorption (EA) of materials. Another goal of the Crash Energy Management Working Group (CEM-WG), of which NHTSA is a member, is to optimize the crashworthiness of light-weight components and vehicle designs. This public-private-partnership research and development effort is congruent with NHTSA's mission interests and could augment NHTSA's PCIV and Hydrogen vehicle safety research.

The DOE-led U.S. Automotive Materials Partnership (USAMP) has undertaken the Materials Technologies Research Program with the goal of "halving personal vehicle weight by using lighter structural materials could result in a 30-percent increase in fuel efficiency"…given that, "as a rule of thumb, for personal vehicles, every 10-percent reduction in weight can result in about a 6-percent increase in fuel economy."[10]

The primary goals of the USAMP are to accelerate development of high-strength, light-weight automotive materials, but "without compromising durability, reliability, and safety," and to achieve the "reduction of vehicle weight without compromising safety."

Several strategies and materials selection options for achieving substantial vehicle weight reductions are under active research to address the National Energy Policy goals and the Energy Policy Act of 2005 mandates. The materials options under investigation are summarized in Figure 1-1. The United States Council for Automotive Research (USCAR) partnerships are examining, in addition to plastics and composites, other alternative materials options for light-weighting, the use of metal alloys, such as aluminum, magnesium, titanium, zinc, and high performance stainless steels (HPSS). This study focuses on reviewing, in general terms, the state of knowledge and research and development priorities related to the safety performance of automotive plastics.

---

[9] CAR stands for Cooperative Automotive Research; also cited in other DOE documents as Council for Automotive Research. See references by J. Carpenter, 2006, and R. Sullivan, 2006.

[10] Figures cited in *Driving Technology: A Transition to Enhance Energy Security,* May 2006.

Weight Savings and Costs for Automotive Lightweighting Materials

Lightweight Material	Material Replaced	Mass Reduction (%)	Relative Cost (per part)*
High Strength Steel	Mild Steel	10-25	1
Aluminum (Al)	Steel, Cast Iron	40 - 60	1.3 - 2
Magnesium	Steel or Cast Iron	60 - 75	1.5 - 2.5
Magnesium	Aluminum	25 - 35	1 - 1.5
Glass FRP Composites	Steel	25 - 35	1 - 1.5
Carbon FRP Composites	Steel	50 - 60	2 - 10+
Al matrix Composites	Steel or Cast Iron	50 - 65	1.5 - 3+
Titanium	Alloy Steel	40 - 55	1.5 - 10+
Stainless Steel	Carbon Steel	20 - 45	1.2 - 1.7

* *Includes both materials and manufacturing.*
Ref: William F. Powers, Advanced Materials and Processes, May 2000, pages 38 – 41.

Figure 1-1: Comparative data on materials substitution options for light-weighting vehicles to improve fuel efficiency (from Carpenter, 2006).

1.2 The American Plastics Council Workshop: Recommendations to Enhance PCIV Safety

Greater use of engineered composites and plastics in automotive applications offers the promise of greater energy efficiency and enhanced safety at affordable cost. However, the potential safety benefits of the next generation of fuel-efficient and environmentally friendly vehicles would have to be planned for, designed, and demonstrated in order to gain market share and public acceptance.

The American Plastics Council (APC, now ACC-PD) and NHTSA co-sponsored a workshop in November 2005 on *Enhancing Future Automotive Safety with Plastics.* The final report (May 2006) summarized the key findings, which are discussed in greater detail in Chapter 3 and listed in Appendix 3.1. The development of a safety research, technology demonstration and evaluation roadmap was recommended in order to enhance automotive safety with plastics, as an integral part of a broader strategy to "light-weight" the automotive fleet for improved fuel efficiency.

The ACC-PD definition of automotive plastics is comprehensive: it includes a broad range of lightweight, strong, and durable engineered materials, including both thermoplastic and thermoset polymer composites with glass, carbon, or other fiber reinforcement embedded in resin matrices and hybrid materials of plastics bonded to metals (sandwich) for added structural strength. When integrated into vehicle skin, structural frame/chassis, power train, interior seats, padding, and/or safety appliances, these plastics must comply with NHTSA's safety standards.

The workshop identified more than 100 recommended priority activities, as well as the critical challenges to overcome in order to enhance future automotive safety, while light-weighting with plastics. ACC-PD also produced a brief Roadmap Prospectus for *Enhancing Future Automotive Safety with Plastics,* which proposed formation of public-private partnership to develop it. ACC-PD estimated that development of a technology integration roadmap would require $1-1.5 million and between 18-30 months to complete. Congress provided NHTSA with a modest amount of research funding ($250,000) in FY06 to initiate research on the potential safety benefits of using more plastics and composites in future PCIVs.

1.3 Project Scope and Technical Approach

In order to develop collaboratively the overall plans and objectives for development of a PCIV Safety Roadmap, and to jointly review interim progress and findings, NHTSA and the Volpe Center hosted regular meetings and maintained continuous telephone informational contacts with ACC-PD technical staff. The Volpe Center focused research project was structured as follows:

- Task 1: Establish Vision for 2020 PCIV Safety - The effort was to develop a vision for desired future (2020) safety, focusing on improved protection of 65 and older drivers with plastics and composites. The vision was based on a review of available DOT and NHTSA Strategic Plans.[11] Also, a review of literature on aging drivers' safety was completed to extract information relevant to future automotive safety needs, knowledge gaps, and research and development priorities. The vision statement for 2020 PCIV safety also identified performance goals and progress metrics.

- Task 2: Situation Analysis - This task was a comprehensive and critical literature review, complemented by interviews of leading automotive safety and plastics industry experts, in order to:

    o Identify the knowledge gaps and research needs, based on industry experience with deployment of automotive plastics and composites in safety applications;
    o Derive lessons learned and best practices to build on success;
    o Identify the most promising NHTSA research and development partnership opportunities for enhancing the automotive safety performance of lightweight vehicles
    o Identify the research, technology and regulatory challenges, and identify strategies for overcoming them; and
    o Recommend consensus research and development priorities for the near-, mid-, and long-term.

---

[11] These documents are currently being updated from the 2005 version discussed in Section 2.1.3, and have not yet been released by NHTSA (as of November 2007). They will be referenced in the final report, if available.

- Task 3: Develop the 2020 PCIV Safety Roadmap - Based on the findings from the situation analysis and on NHTSA and ACC-PD recommendations, the top three most promising safety enhancements will be selected for the Safety Technology Integration Roadmap. Since the ACC-PD 2002 Technology Roadmap did not explicitly address automotive safety, this effort complements and extends previous industry and government strategic plans, which identified enabling research needs for PCIV technology commercialization. This PCIV Safety Roadmap will focus only on the top few applicable safety research and development priorities for PCIVs. Particular attention will be paid to protecting the vulnerable and growing demographic segment of older occupants. This report should identify strategies for improving occupants' crash protection with plastics in safety applications by 2020. It should link the identified safety research and development needed to bridge existing knowledge gaps to PCIV deployment opportunities.

  The analysis and resulting Safety Roadmap for PCIVs will also focus on multiple safety research and development priorities identified. An ultimate goal of the road-mapping process is to better align research within industry, academia, and government. It provides a technical transitional approach that informs decisions regarding public-private partnership R&D investments, as well as serve as a useful communications tool throughout the automotive safety value chain.

- Task 4: Engage the Stakeholders - Automotive safety stakeholders will be engaged in this process and will give their support for goal-oriented research and development and technology applications. Within the limits of available resources, Volpe Center staff will engage a representative cross-section of stakeholders through focused telephone interviews, e-mail correspondence, conference calls, and selected meetings (such as attendance at the annual conferences of the Society of Automotive Engineers (SAE) and the Society of Plastics Engineers (SPE) Automotive Division- ACCE06 and 07). Safety technology stakeholders included: leading technical experts on crashworthiness, plastics, and composites from federal agencies, universities, automotive Original Equipment Manufacturers (OEMs), insurers and representative advocacy groups for automotive safety and for the aging.

2. A 2020 VISION FOR PCIV SAFETY

2.1 Resources for a 2020 PCIV Safety Vision Statement

2.1.1 The Department of Transportation (DOT) Strategic Plan, 2006-2011

The DOT Strategic Plan for 2006-2011 Safety Strategic Goal[12] adopts the explicit strategy **"to enhance vehicle safety through the introduction of new technology and to assess the lifesaving benefits of emerging technologies as they enter the vehicle fleet."**

The behavioral safety goals for reducing the number of crashes and mitigating the severity of their consequences include:
- "increasing occupant protection;" and
- "extending the mobility of older drivers for as long as medically practicable."

Another explicit strategy is to **"Partner with key stakeholders to promote the use of engineering design features that reduce crashes."** This would encourage NHTSA to develop a new research and development partnership initiative on PCIV research with other Federal agencies, other DOT modal Administrations, industry, and public-private research consortia.

The desired *outcome* is to decrease the rate of fatalities and injuries in motor vehicle crashes. The *performance measure* for this outcome is the annual number of highway fatalities and injuries per 100 million vehicle miles traveled (VMT).

The PCIV vision, goals and performance measures need to be consistent with these DOT and NHTSA highway safety goals and performance metrics.

2.1.2 The DOT Strategic Research Plan

The Transportation Research, Development and Technology Strategic Plan 2006-2010[13] crosscutting research plan specifies two quantitative safety goals for highway fatalities in 2011, which NHTSA, FMCSA, FTA, and FHWA would have to address cooperatively and contribute to in a quantifiable manner:

- The 2011 target for highway fatalities per 100 million VMT is about one (1.0).
- The 2011 target for highway fatalities involving large trucks per 100 million truck vehicle miles traveled (TVMT) is 1.65.

The PCIV Safety Roadmap research project is one of the three NHTSA research initiatives identified in the plan in support of the research, development, and technology (RD&T) strategy

---

[12] The 2006-2011 DOT Strategic Plan is posted at http://stratplan.dot.gov

[13] The November 2006 plan is posted on the Research and Innovative Technology Administration (RITA) Web site at http://rita.dot.gov/publications/transportation_rd_t_strategic_plan/html

for the safety goal, namely to "Assess Impacts of New Technologies, Vehicles, Concepts, Designs, and Procedures."

2.1.3 NHTSA Vehicle Safety Research Plans and Programs

The NHTSA Vehicle Safety Rulemaking and Supporting Research Priorities for CY 2005-2009[14] outline the NHTSA safety regulatory development plans and respective supporting research programs. Several research and development priorities are synergistic with the goal of enabling the future deployment of fuel-efficient and safe PCIVs. Such ongoing NHTSA safety R&D initiatives include efforts to:
- address crash compatibility challenges (both self- and partner-protection);
- prevent and mitigate rollover crashes and reduce occupant ejections;
- develop advanced dummies, injury criteria, and performance levels; and
- better protect older drivers and occupants in crashes.

The FY07 NHTSA Congressional Budget Request stated that the desired safety outcome is to **reach one fatality per 100 million VMT by 2008 (from 1.44 in 2004)**. This is equivalent to a drop in total fatalities from 46,700 to 32,000; in spite of the projected steady growth in vehicle miles traveled (VMT) and the number of licensed drivers.

The *NHTSA Budget Overview FY07* also stated that a safety priority is to "improve motor vehicle crashworthiness, crash protection, and injury mitigation," which could be achieved by using advanced occupant protection and crash avoidance technologies. Although the NHTSA research and development program has not explicitly addressed advanced materials research and development, several research priorities for future PCIV vehicle safety (further discussed in Chapter 4) are compatible with NHTSA research program priorities and strategies.

Contingent on continued funding by the Congress for exploring potential PCIV safety benefits, closer integration and synergies with related NHTSA Safety Systems research programs could be developed. Such related research topics include: advanced air bags, improved safety belt and head restraints, frontal crash protection, ejection mitigation or reduction, side crash and roof crush protection and rollover mitigation. Improved vehicle-to-vehicle compatibility in collisions is a strategy being addressed by NHTSA research and by industry initiatives.[15]

Innovative strategies to ensure occupant crash protection for all vehicles, including PCIV collisions, with heavier and larger vehicles, will have to be developed and demonstrated. The NHTSA 2004 CAFE Rule[16] also discussed the impact of vehicle weight reduction strategies on crash safety and survivability, and the statistical evidence for linking safety to vehicle size,

---

[14] The 2005 document, which is currently in update and revision, is posted at http://www.nhtsa.gov/cars/rules/rulings/PriorityPlan-2005.html and a summary for Congress is available at http://www.nhtsa.dot.gov/nhtsa/announce/NHTSAReports/PriorityPlan-2005.html

[15] Patel, S., Smith, D., and Prasad, A. (2007). *NHTSA's Recent Vehicle Crash Test Program on Compatibility in Front-to-Front Impacts*. International Conference on the Enhanced Safety of Vehicles (ESV).

[16] 49CFR Parts 523, 533 and 537, Average Fuel Economy Standards for Light Trucks Model Years 2008-2011 (2004), pp. 166-167 and 189-196.

height, weight, and technologies. NHTSA's analysis concluded that "… design changes can also reduce a vehicle's weight without reducing crashworthiness, and may in some instances improve the safety of a vehicle (e.g., reduce rollover propensity)." This safety performance challenge must be met by all future motor vehicles, including PCIV candidate designs.

Other safety research issues to enable future PCIV deployment are beyond the scope of this project. The main thrust of NHTSA's safety research is the PCIV structural strength performance, optimal load distribution or force attenuation (cushioning) in safety devices, and crash energy management using "crush boxes" (also called "crumple zones"). Other safety considerations for PCIVs might include material damage tolerance and maintainability; flammability; the durability of plastics when exposed to environmental stressors (temperature, UV and solar heat, corrosion); and the potential toxicity of volatile organic compounds (VOC) released at higher environmental temperatures and/or as byproducts of crash-related fires. Although some reports[17] have claimed that plastics in cars (when new or heated) outgas toxic volatile compounds, recent research[18] has concluded that there are no apparent air quality health hazards due to VOC levels measured in parked motor vehicles.

2.1.4 The NHTSA Integrated Vehicle Safety Research Strategy

The new strategic approach to integrated vehicle safety research was outlined in public presentations by research and development leadership in the past year. A recent presentation by Dr. William T. Hollowell[19] articulates in the NHTSA Strategic Plan Goal #1 – the new integrative and crosscutting systems approach to research and safety regulatory development, as illustrated in Figs. 2-1 and 2-2.

---

[17] See report *Toxic at Any Speed: Chemicals in Cars and the Need for Safe Alternatives* posted at www.healthycar.org and www.ecocenter.org.

[18] See referenced paper by Buters et al. *Toxicity of Parked Motor Vehicle Indoor Air.* Environ. Sci. Technol. 2007, 41, 2622.

[19] November 10, 2006, presentation to the MADYMO International Users Meeting, entitled *Overview of NHTSA Research for Enhancing Safety.*

Research Mission and Goals

- **Advance Scientific Knowledge in Support of NHTSA's Mission**
  - Identify Opportunities for Increasing Safety
    - Crashworthiness, Crash Avoidance, Restraint Systems
  - Understand Performance of Advanced Technologies
    - Crash Prevention, Crash Severity Reduction, Integrated Safety
  - Develop Objective Tests to Discriminate Performance
  - Address Human-Vehicle Perfomance Issues
  - Retain Leadership in Human Injury Research
  - Support Rulemaking, Consumer Information and Enforcement Programs Research Programs in Vehicle Safety

Figure 2-1: NHTSA Strategic R&D programs support its safety regulatory mission.

Figure 2-2: NHTSA Vision Statement (W.T. Hollowell "Overview of NHTSA Research for Enhancing Safety," November 10, 2006, Presentation to MADYMO International Users Meeting in Detroit, Michigan.)

2.2 The need to enhance the safety of future vehicles and older driver protection

2.2.1 Crash protection needs of older drivers and occupants

The development of the vision statement for PCIV Safety and the definition of associated targets and milestones, as well as the justifications of need that are documented below are based primarily on the NHTSA and ACC-PD safety priorities. They were, however, well supported by the Situation Analysis literature review as summarized in Chapter 3, and by the experts' inputs discussed in Chapter 4. The 2006 U.S. Census report *65+ in the United States: 2005* projected that the aging "baby boomers" will lead to a sizeable increase of the 65 and older population segment as shown in Figure 2-3 below:

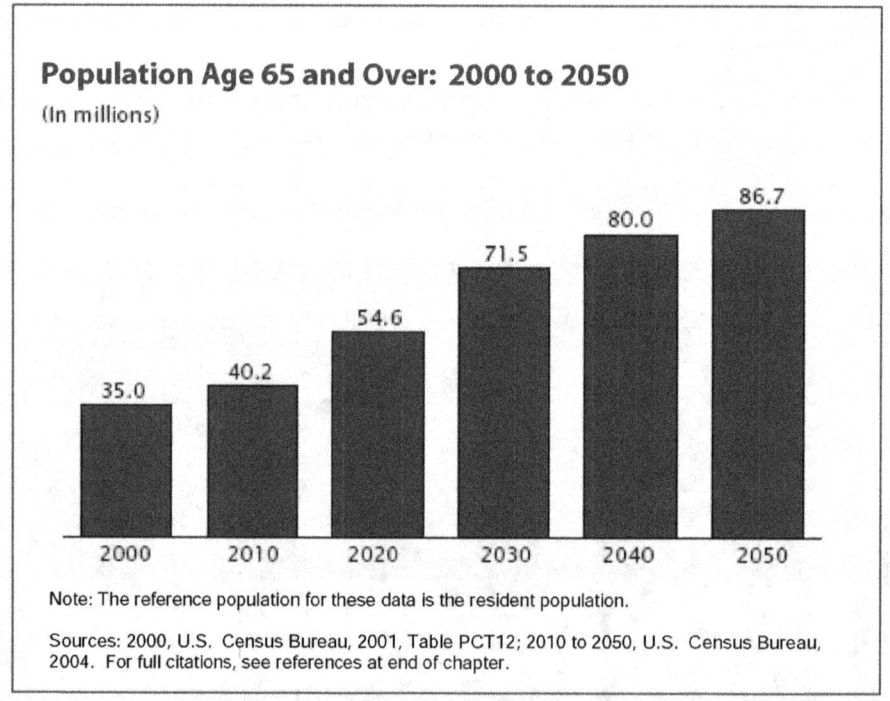

Figure 2-3: From the U.S. Bureau of Census special report *65+ in the United States: 2005*

Currently, **the 12.5 percent of the U.S. population over age 65 accounts for almost one-third of all deaths from injury and incurs a higher population-based death rate than any other age group**. Motor vehicle crashes are the most common reason for the elderly to be transported to a trauma center. Several large studies have shown **that despite driving less overall, the crash rate per mile driven takes a sharp upward increase after age 60.** The public health impact of this upward trend is further compounded by the fact that **not only are older occupants likely to sustain more serious injury in any given motor vehicle crash, but the risk to life is significantly greater for a given injury severity. An improved understanding of the mechanisms and patterns of traumatic injuries sustained by the elderly in motor vehicle crashes is essential if better preventive countermeasures are to be devised.**
Reference: *"An Aging Population: Fragile, Handle with Care"*, Stewart C. Wang, UMTC (from the NHTSA CIREN Web site)

Adults 60 and older, so-called "baby boomers," represent the most rapidly growing demographic segment, as shown in Figure 2-4. The 2006 Census Bureau data[20] projected that the "over 65" of the population is expected to nearly double over the next 25 years (by 2030), from 12 percent to 21 percent and reach 71.5 million. This major demographic shift could potentially create road safety concerns by 2020 and beyond, in view of their higher crash rates and their greater vulnerability to injuries and fatalities in crashes. The future safety challenges for an aging population are likely to rise, in view of the fact that drivers in their late seventies have a triple fatality rate per VMT than drivers aged 30-65. The risk is even higher for drivers in their eighties.[21]

Balancing the elderly transportation safety and mobility needs is indeed a global concern and the subject of numerous studies.[22] An important objective is to improve vehicle safety, although this is only one of multiple factors, such as: aging drivers' screening and education, improved visibility of signalization and road infrastructure, and transportation alternatives on demand. Strategies to enhance vehicle safety include not only structural crashworthiness improvements, but also use of multiple adaptive occupant protection devices, and of advanced Intelligent Transportation Systems (ITS) consistent with NHTSA's integrated safety approach.

The ACC-PD workshop summary stated that more rapid deployment of plastics materials in automotive safety applications promises[23] to address the highway safety needs of aging baby-boomers as a growing demographic segment. The review of relevant literature reveals the following demographic trends and related safety challenges and opportunities for future PCIVs:

- In 2006, about 12.4 percent of the population (or 1 in 8) is 65 and older, but the estimated proportion by 2030 will be 20-25 percent.
- The number of people 65 and older is expected to double to 71.5 million by 2030.
- An American Association for Retired Persons (AARP) fact sheet indicates that in 2003, a high proportion (about 80%) of people older than 65 are licensed drivers. About 1 in 7 licensed drivers is 65 or older.
- The Insurance Institute for Highway Safety (IIHS)[24] projected that, if current trends hold, by 2030 the drivers over 65 will represent 25 percent of all drivers involved in crashes, and 16 percent of drivers in fatal crashes, compared to <10 percent at present.
- Fragility (as related to the risk of death following a crash) is the primary reason for the elevated risk per unit exposure for older drivers and passengers.

---

[20] See American Fact Finder Census data on Aging and Age and Sex posted at http://factfinder.census.gov/servlet/ and Elderly (65+) population statistics posted at http://www.census.gov/population/www/socdemo/age.html#elderly

[21] Shane, J.N., Under Secretary of Transportation for Policy. *The Surface Transportation System: Challenges for the Future.* Testimony before the Committee on Transportation and Infrastructure, Subcommittee on Highways and Transit, US House, January 24, 2007.

[22] See TRB Conference Proceedings, *Safe Mobility for Older Americans* (2005); *Transportation in an Aging Society* (2004); and *The Elderly and Mobility: A Review of the Literature,* Report 255 (2006). Monash University.

[23] American Plastics Council (APC). (2006). *Enhancing Future Automotive Safety with Plastics: Technology Integration Report.*

[24] IIHS Status Report Special Issue: Older Drivers Volume 42, No. 3, March 19, 2007.

NHTSA crash statistics for drivers over 70, shown in Figure 2-4, indicate that the total number of fatalities for older drivers in crashes have increased over time, in spite of decreasing death rates over time due to vehicle safety improvements. This is partly due to increasing fragility, degradation of vision, health, and reaction time of aging drivers. The crash safety situation for older drivers and passengers as summarized in the text box below holds true today and will have to be addressed in the future:

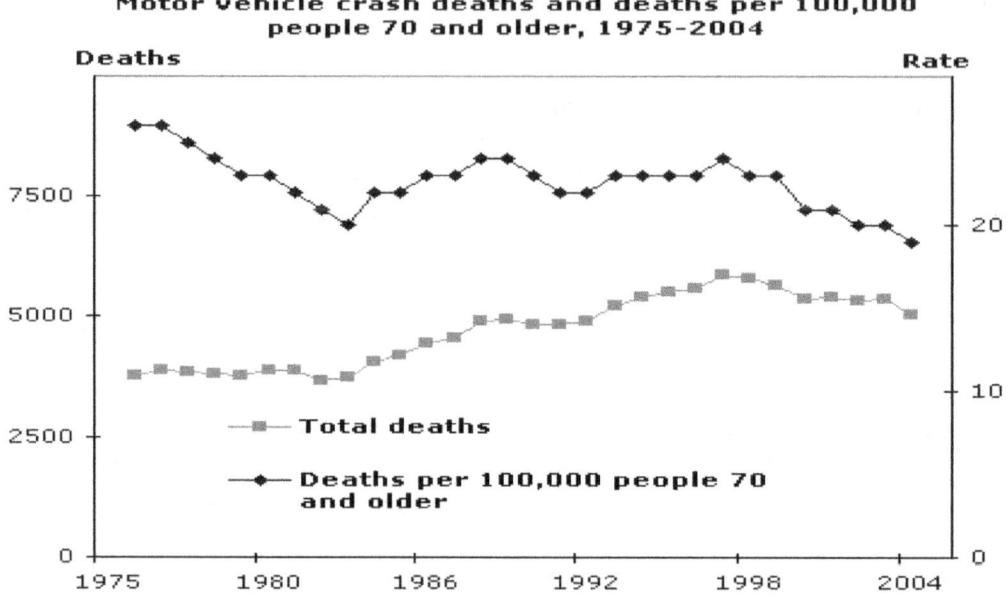

Figure 2-4: Total fatalities and normalized rates for the over 70 drivers versus time.

As shown by the MIT Age Lab data forecast in Figure 2-6, although crash safety has been improving over several years, older drivers are still at higher risk of fatality and injury because of the forecasted VMT growth associated with the increase in aging drivers on the road. The 2001 Insurance Institute for Highway Safety (IIHS) analysis and summary of the situation cited below is still valid:

> This status report found that the fatality crash rate per vehicle miles traveled (VMT) is highest for both older and younger drivers. The reason for the higher risk of death of older drivers is the increasing body fragility with age, not because they are involved in more crashes. Improved vehicle and road changes and changes in driving behavior could improve the crash survivability of older drivers. To make the driving environment safer and more comfortable for older drivers, "improved occupant protection is needed including: vehicle design changes (such as installing belt force limiters, improving safety belt systems, and advanced air bag technology), as well as improved vehicle ergonomics, road visibility aids, and active crash avoidance technologies."
>
> Reference: ***Older Drivers Up Close Aren't Dangerous Except Maybe to Themselves, Status Report, Insurance Institute for Highway Safety, Volume 36, No. 8 (2001).***

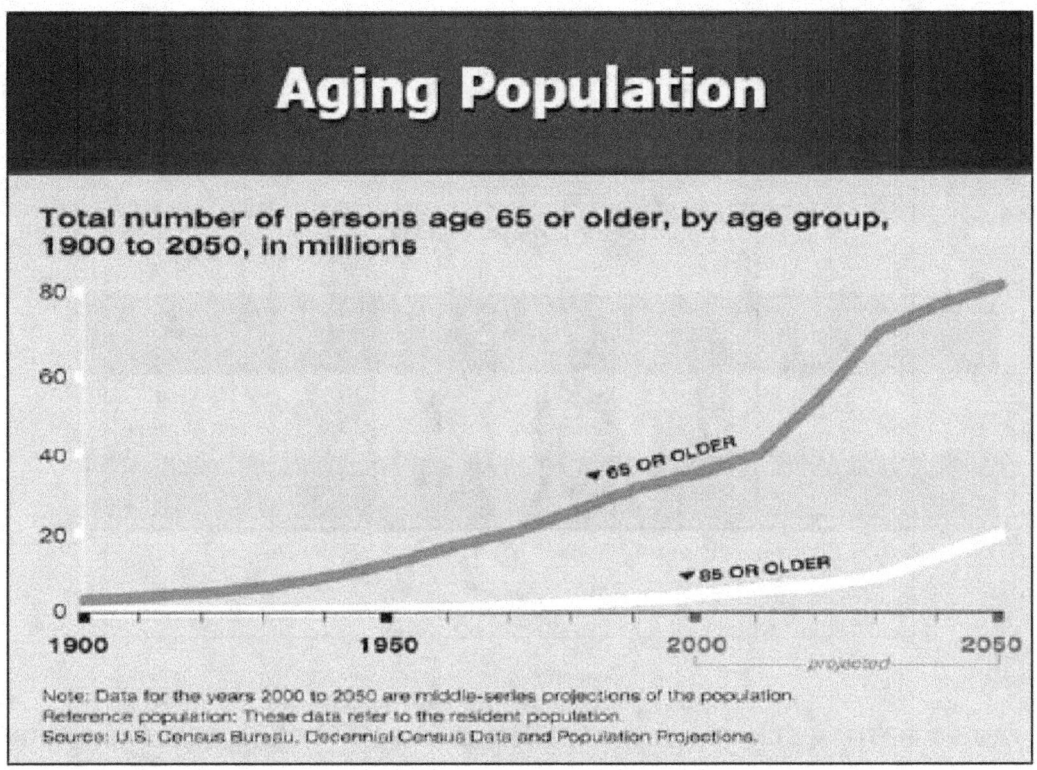

Figure 2-5: Projected totals of aging "baby boomers" by age group, to 2050.

Because the life expectancy would also continue to grow as medical care improves, the ranks of aging drivers will swell, with geographic concentration in sunbelt States.[25] As the older drivers' health and driving skills decline with age (impaired vision acuity, reaction time) and their bones become more fragile, the threshold for injury and fatality in automobile crashes decreases and the severity of injuries increases. The Crash Injury Research and Engineering Network (CIREN) data indicate and crash safety experts argued that upgrades in torso and head protection for elderly occupants in side impacts are a pressing need, in view of their lower harm thresholds and higher vulnerability to injury trauma.[26]

Older drivers today also tend to drive older model cars, which have fewer and possibly outdated safety features. However, since the personal vehicle fleet turnover time is about 15-20 years[27], the elderly are expected to drive much safer cars by 2020.

---

[25] Ref: TCRP Report 82 *Improving Transportation Options for Older Persons* (2002) by Westat, Inc. and *Mobility and Independence: Changes and Challenges for Older Drivers* by Ecosometrics, Inc. for NHTSA.

[26] Joan Claybrook, President, Public Citizen, *Comments on Side Impact Protection NPRM 69CFR27990, May 17, 2004,* Oct 14, 2004.

[27] See Mintz et al. (2000). *From Here to Efficiency: Time Lags Between Introduction of Technology and Achievement of Fuel Savings* (Transportation Research Record 1738), p.100-105, TRB, TRISonline.

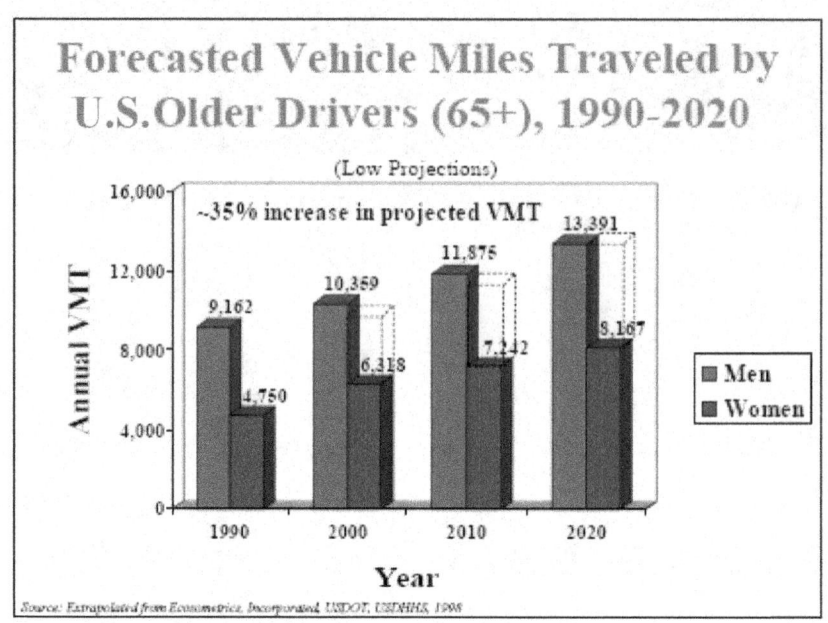

Joseph F. Coughlin
Age Lab - Massachusetts Institute of Technology
Figure 2-6: MIT Age Lab forecasts a 35-percent increase in VMT for aging drivers by 2020.

The elderly are less familiar and comfortable with modern electronic active safety devices, which may require training and might even distract them. Therefore, there is a need for rapid deployment of passive safety systems, along with training in the use of well-designed, integrated and user-friendly active safety and driver-assisted devices. Emerging technologies that promise to further enhance elderly safety, but may require training, include: haptic and/or audible lane-change or collision warning and avoidance systems, night-vision aids, and heads-up-display components (see references by Coughlin et al.).

Passive and active occupant protection systems are designed to protect all occupants, as are the ITS technologies being developed under the Integrated Vehicle Based Safety Systems (IVBSS) Program.[28] Several emerging automotive safety technologies, if implemented in future PCIVs, could provide age-differentiated occupant protection based on driver's and passenger's weight, size, and fragility: multi-point belt systems, seat belt force limiters, advanced (smart) air bags with staged deployments geared to impact force, side air curtains and leg/knee bolsters, ergonomically designed enhanced seating systems, visual displays and auditory alerts, and the application in commercial vehicles of racecar safety technologies, such as head and neck restraint systems (HANS) and padded body harness.

In view of the clear need to substantially enhance crash protection for vehicle occupants, which includes the aging, a major focus of a PCIV safety research and development roadmap would be to reduce the number of crash fatalities, and to improve the injury survivability. The integrated safety approach adopted by NHTSA is designed to enhance the safety of all next generation

---

[28] See presentations posted at http://www.itsa.org/ivbss.html

vehicles, including PCIVs. Therefore, the introduction of advanced materials and technologies for improved structural crashworthiness and for interior protection appears promising, especially when and if synergistically integrated with advanced occupant restraints and active safety systems such as Electronic Stability Control (ESC).[29]

Research is necessary to determine what role various plastics and composites materials could play in proposed safety applications for 2020 PCIV designs, relative to other light-weighting materials options and safety technologies. Data needs include:
- Definition of safety performance criteria for 2020 vehicles, sufficient to protect all occupants, including the elderly;
- Age-adjusted thresholds for injury criteria;
- Evaluation of potential uses of plastics in structural or safety applications so as to demonstrably improve the crash safety for all occupants;
- Determination of quantitative or qualitative tests to demonstrate the PCIV safety improvements over time;
- An attribution methodology, to quantify and evaluate specific contributions of plastics to the integrated vehicle safety, separately and due to the simultaneous application of active and passive safety technologies, e.g., Electronic Stability Control (ESC), rollover roof air bags, variable ride-height suspension (VRHS), and other innovations for vehicle crash energy management, crash avoidance, and night visibility enhancements.

There is a need to clarify how realistic and pertinent it is to set a quantitative goal for reduction of the current fatality and injury rates for older drivers by 2020, and to better understand the complex interplay of multiple factors and trends, e.g.:

- The 2020 fleet makeup, with evolving types of vehicles, including PCIVs;
- The total number of older drivers on the roads, and their normalized risk exposure in terms of VMT for 2020;
- The regulated safety features on all cars, including PCIVs, in 2020;
- Their safety performance relative to that of current vehicles;
- Modeling and simulation tools and specific crash tests able to fairly credit and reveal the contribution of specific advanced materials and devices to the overall safety performance of the vehicle.

2.2.2 The need to capture the potential safety benefits of PCIVs

Improved vehicle-to-vehicle compatibility in crashes is a key safety issue for all vehicles, especially if PCIVs coexist with heavier and larger vehicles. It is also a priority focus of ongoing NHTSA and industry research to enhance vehicle safety and improve crash compatibility.

Currently, the safety performance in crashes for the smaller and more fuel-efficient lighter-weight vehicles is considerably worse than that of heavier vehicles. The most recent (Dec. 2006)

---

[29] The NHTSA Final Rule for FMVSS 126 (issued in March 2007) requires ESC for all vehicles under 10,000 lbs. to be phased in starting with MY 2009 by 2012.

IIHS crash-test data[30] indicate that the most fuel-efficient mini-cars, subcompact, and compact cars **have twice the fatality rates of heavier cars,** as shown in Figure 2-7.

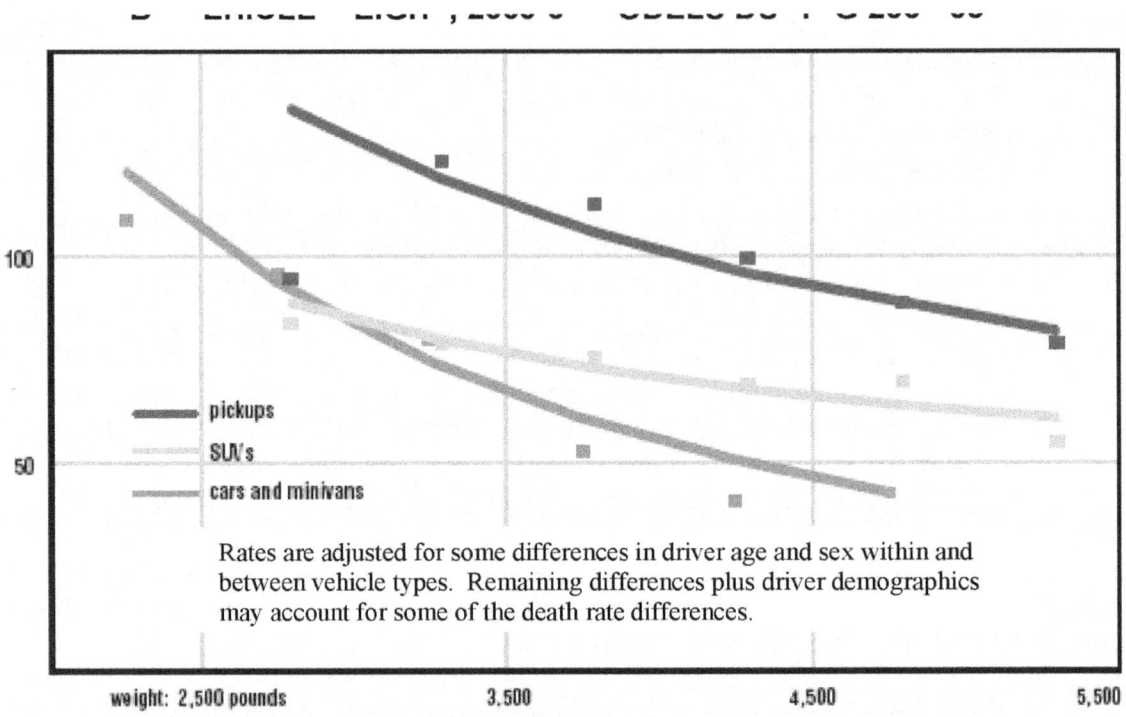

Figure 2-7: From the December 19, 2006, IIHS Status Report newsletter, "Bigger is generally better" shows that the risk of crash fatality is higher for lighter, more fuel-efficient vehicles at present.

Analyses of the crash-safety implications of the relative vehicle weight, size, fuel efficiency, and other design parameters (vehicle height, geometry) and factors (driver aggressivity) have been discussed in the literature.[31] [32] For instance, Ahmad and Greene's updated 2004 analysis of correlations between fuel economy and safety for passenger cars show that higher mpg are actually correlated with fewer fatalities. Some presenters at a recent Experts Workshop[33] argued

---

[30] Insurance Institute for Highway Safety (IIHS). News Release, December 19, 2006: *First Crash tests of minicars,* and Status Report (Vol. 41, No. 10) with crash safety ratings for five 2006 and 2007 minicar models, posted at www.iihs.org.

[31] See listed references by NHTSA's Kahane, and by Evans, Lovins, Ahmad and Greene, and Ross et al.

[32] Gordon, D., Greene, D., Ross, M., and Wenzel, T. (2007). *Sipping Fuel and Saving Lives: Increasing Fuel Economy Without Sacrificing Safety.* Report by October 2006 experts workshop "Simultaneously Improving Vehicle Safety and Fuel Economy Through Improvements in Vehicle Design and Materials."

[33] *Sipping Fuel and Saving Lives: Increasing Fuel Economy Without Sacrificing Safety.* Summary by Deborah Gordon, David L. Greene, Marc H. Ross, and Tom P. Wenzel of an October 3, 2006, Experts Workshop on Simultaneously Improving Vehicle Safety and Fuel Economy through Improvements in Vehicle Design and Materials, posted at
http://www.hewlett.org/Programs/Environment/Energy/Publications/Sipping+Fuel+and+Saving+Lives.htm

that the "laws of physics" should not be invoked to the disadvantage of lighter and smaller cars in collisions; and that in spite of historic crash data, "correlation is not causation." Others attributed safety disadvantages of smaller cars to the fact that historically, the smaller, lighter, and cheaper cars have not been equipped with the best passive and active protection systems. The workshop summary cited identified future vehicle designs, materials, and technologies that could simultaneously ensure safety and improve fuel efficiency.

Given the superior structural strength of composites, it was argued that lightweight cars could be designed to be both safe and fuel-efficient, especially if size can be decoupled from weight factors (for instance by allowing composite crush cones in front and back to make a car larger, yet lighter). Furthermore, if all modern passive and active safety appliances were provided to compensate for their lower momentum and kinetic energy in crashes of smaller and lighter cars, they could effectively protect the occupants. Future 2020 lightweight vehicle designs will have to demonstrate a high standard of safety performance in order to be accepted by the public and obtain sizeable market share.

2.3 Vision Statement for PCIV Safety

2.3.1 Resources for PCIV Safety Vision

A key objective of this project was to develop a 2020 Vision statement for enhancing automotive safety with increased use of plastics and composites, with special focus on improving crash protection for elderly drivers and passengers. An associated goal was to identify safety performance metrics and targets for the near-term (3-5 years), mid-term (5-10 years), and long-term (10-15 years).

The proposed Vision Statement for 2020 PCIV safety performance considers simultaneously the most important ACC PD safety activities, the DOT Strategic Plan and NHTSA research and development and regulatory priorities, and the future safety challenges discussed above. Recent DOT and NHTSA strategic planning and budget documents (listed in Section 3.1 and in Chapter 6) were used to identify and document both the NHTSA policy and research objectives to enhance automotive safety and efficiency, and clarify their relevance to technical goals for plastic and composite vehicles. The APC vision (from the 2002 Technology Plan) and automotive industry vision for future PCIVs (from SPE Automotive Division annual conferences and panels) were also considered.

The proposed 2020 Vision Statement for ensuring safety benefits of lightweight PCIVs is a "stretch goal," with interim safety performance targets and milestones suggested for the near-term (3-5 years), mid-term (5-10 years) and long-term (10-15 years) to 2020. The proposed Vision Statement describes a desired future state of automotive design and crash safety performance, which can be accomplished not only through policies, funding promoting cooperative research and technology "push," but also through free market competitive "pull" on plastics suppliers and the automotive industry.

> **VISION STATEMENT FOR PCIV SAFETY IN 2020:**
> *"NHTSA—in partnership with other government agencies, industry, and academia—will support research on safety-centered design and performance modeling and validation to enable and foster superior, integrated safety performance of future light-weight Plastics and Composite Intensive Vehicles (PCIVs)."*

This vision statement reflects the 2006 Congressional requirements, and is consistent with the NHTSA "integrated vehicle safety" mission and with APC workshop recommendations. This vision statement, with associated mission activities, and measurable performance goals, can:

- Support the safety strategic goals and objectives identified in the NHTSA Strategic Research Plan and Regulatory Development Plan.
- Exploit synergies with the ACC-PD and national Department of Energy research and development objectives of improving the fuel efficiency of the vehicle fleet through weight reductions, while preserving or improving their safety performance.

2.3.2 PCIV Safety Research Goals and Objectives

The supporting goals and objectives for this PCIV 2020 Vision are to:

- Conduct cooperative research and development with government, industry, and academia, to promote the development and deployment of advanced safety solutions in future light-weight PCIVs;

- Monitor and measure progress towards demonstrating and enhancing the safety performance of prototype PCIVs;

- Review and evaluate applicable NHTSA safety regulations and consumer information and education programs. NHTSA's interest is to ensure that they are objective, materials-neutral, and performance-based.

- Participate in the development of voluntary technical consensus safety standards for vehicle crashworthiness and related materials testing. This would facilitate the development and deployment of innovative plastics or other advanced materials in automotive safety applications.

2.3.3 Performance Metrics and Milestones

In order to measure progress towards achieving the 2020 PCIV Safety Vision, performance metrics and milestones should be defined. For an applied research program, such performance metrics, or indicators, frequently refer to intermediate outputs rather than an outcome. The outcome would not be achieved until the PCIVs are deployed as a sizeable component of the vehicle fleet. Plausible performance indicators for a future NHTSA cooperative research and development program on PCIV Safety (if authorized and funded by Congress) could include:

- Development of improved predictive engineering modeling and simulation tools for PCIV crashworthiness that can be then compared to actual crash testing performance in a verification and validation program (near- and mid-term, by 2015);
- Research and development projects that monitor, evaluate, contribute to, and integrate the knowledge base on mechanical properties and the crashworthiness of composite materials in PCIVs (near- and mid-term, by 2015);
- Development and application of advanced crash modeling and simulation tools to predict PCIV safety performance;
- Support of professional and scientific societies for the development of standardized materials testing and best engineering practices for verification and validation of PCIV crashworthiness (mid- and long-term).
- Periodic regulatory reviews to ensure a level playing field for PCIV crash safety performance and for alternative vehicle designs and materials. As discussed above, this would be accomplished in partnership with professional associations, government, and industry (mid- to long-term).

In each case, the safety performance metric could be any quantitatively demonstrated improvement in occupant safety, benchmarked every five years up to 2020, for lightweight PCIVs, such as the decrease in normalized fatality and injury rates for occupants, relative to the 2006 baseline:

- The fractional reduction in fatality rate per million VMT, for all drivers;
- The fractional decrease in the crash injury severity for all occupants and the corresponding increase in their survivability; and
- Reduction in the disparity in crash injury severity between average adults and older occupants.

As discussed in Chapter 3, this vision statement, goals, and objectives for a new NHTSA research and development foundational initiative to improve PCIV safety are consistent with other ongoing DOT multi-modal research and development programs, such as the FAA Aircraft Composites Safety and Certification Initiative; the FHWA advanced composites for infrastructure applications and for highway safety appliances; and the FTA composite bus research, development, and demonstration effort. Related goals and objectives are suggested in Chapter 4 by the experts who provided inputs to this study.

3. SITUATION ANALYSIS—APPROACH AND FINDINGS

3.1 Technical Approach

The Volpe Center conducted a comprehensive "Situation Analysis" to define the most promising automotive safety applications of plastics and composites, and to identify appropriate safety enhancements of future PCIVs in general, as well as to improve protection of aging drivers and occupants in particular. This effort included several key environmental scan components:

- The ACC-PD safety research and technology integration priorities (Sec. 3.2): A review, analysis and categorization of the ACC-PD. Workshop findings and recommended safety enhancements for future plastics and composite intensive vehicles (PCIVs).

- Literature review: A comprehensive review of the technical and trade literature on automotive plastics and composites and of conference proceedings (e.g., the ACCE06 and preceding conferences of the Society of Plastics Engineers Automotive Division). A comprehensive set of references was critically reviewed (see Chapter 7: References), including those provided by the experts surveyed.

- Older Drivers issues: A review of literature concerning the demographic trends and special safety issues for elderly drivers and passengers, which should be addressed to enhance their protection in general, with future PCIV safety features, design, and materials.

- Existing Research Partnerships (Sec. 3.4): A critical review and evaluation of ongoing Federal agency R&D programs was conducted, including university-based and public-private-partnerships (P3), in order to identify:

    o Technical accomplishments relevant to the crashworthiness of automotive plastics and composites;
    o Industry Best Practices (BP) and "lessons learned" from the Partnership for a New Generation of Vehicles (PNGV) and FreedomCAR and fuel programs;
    o Potential partners, processes, timetables, and procedures that could enable the successful integration of automotive plastics with direct safety benefits;
    o Ongoing industry and Federal R&D and demonstration P3 efforts which offer high-leverage opportunities for NHTSA safety research, including but not limited to: the previous PNGV, the National Institute of Standards and Technology (NIST) Advanced Technology Program (ATP) and Manufacturing Extension Program (MEP); the National Science Foundation program on Predictive Engineering and Advanced Materials; and primarily, as Congress directed, the Department of Energy and industry research and development partnerships (within FreedomCAR conducted by the U.S. Council for Automotive Research);

- o Specific ultra-light and ultra-strong materials proven to contribute to the crashworthiness and driver safety of racing cars, high-end sports roadsters (supercars) and industry prototypes, which could be commercially deployed in mass-market commercial vehicles over the next 15 years; and
- o Concept cars (like those shown at the 2007 North American International Auto Show) and emerging vehicle designs and prototypes (like the RMI "hyperlight hypercar").

- <u>Safety Standards and Guidelines for Automotive Plastics and Composites</u> (Sec. 3.4): A partial survey of existing and draft standards relevant to the crashworthiness of automotive components, either existing or under development by professional standards organizations and trade associations; Society of Automotive Engineers (SAE), American Society for Testing Materials (ASTM), Materials Research Society (MRS), and the Society of Plastics Engineers (SPE).

- <u>Advocacy groups</u>: This entailed review of relevant publications by non-governmental organizations with highway safety advocacy focus: publications and Congressional testimony from non-governmental advocacy groups of stakeholders (e.g., Insurance Institute for Highway Safety [IIHS], the American Association for Retired Persons [AARP], Consumer Reports, American Automobile Association [AAA], Advocates for Highway and Auto Safety, Science Serving Society, Public Citizen, and Environmental Defense), regarding the crash safety performance of lightweight vehicles and enhancement needs, and the safety issues of elderly drivers and passengers.

- <u>Survey of Experts</u>: Chapter 4 discusses the process, inputs, consensus research and development priorities, and challenges for PCIV safety provided by a focused Delphi survey of leading experts on automotive plastics in structural or occupant safety applications. They included representatives from government, industry and academia, suggested by ACC-PD, NHTSA, and the Department of Energy. They responded to our structured interview guide by phone and/or e-mail. The experts who responded focused their inputs on PCIV safety research needs, priorities, and strategies to enhance future PCIV safety performance and assist in identifying priority consensus issues. The structured interviews of experts facilitated the identification of consensus research and development priority needs in the near-, mid-, and long-term for the Roadmap development effort.

3.2 The ACC-PD Safety Priorities for Future PCIVs

Over 100 priority technology roadmap activities to enhance future automotive safety with plastics were identified in the ACC-PD workshop and report, and are summarized by category in Appendix 3.1. Table 3-1 identifies, consolidates, and categorizes the most promising priorities for further development in a safety technology integration Roadmap, which are most closely related to the NHTSA safety mission.

The ACC-PD is engaged in outreach and education to promote increased utilization, diverse applications, and more rapid technology integration of automotive plastics and composites.

Table 3.1: ACC-PD Priority Technology Integration Activities

Priority Activity for Improved Safety
1. Body and exterior
Improve energy absorption in frontal and side impacts
Rollover roof crash performance; and
Crash Energy Mgt. (CEM) for occupant protection in frontal, side, and rear impacts
Develop test standards and performance specifications.
2. Interior
Master Plan for test standards, materials classification and modeling of plastics
Improve Vehicle-to-Vehicle Compatibility in side impacts crash performance
Improve crash energy management with plastics (belts, air bags, foam structures)
3. Powertrain and chassis
Optimize safety and fuel efficiency
Improve predictive modeling for polymer composites
Develop and validate non-flammable laminates and sandwich structures (metal-plastic hybrids)
4. System light-weighting
Characterize safety performance of plastics in partnership with NIST, Department of Energy labs, ACC-PD, USCAR using: Materials Database, Models and Simulations, Coupon, Component and System Level Crash Tests
Standardize materials performance specs in crashes and safety test integrated panels/components
Active and passive integrative safety strategies

The ACC-PD has an ongoing Cooperative Research and Development Agreement with the Department of Energy Office of Energy Efficiency and Renewable Energy (DOE/EERE)/Argonne National Lab (ANL) and the Vehicle Recycling Partnership of USCAR, to develop recycling technologies for end-of-life vehicles of the future and explore recyclable automotive plastic components. ACC-PD is also working with other Department of Energy labs (PNNL and ORNL) and the National Science Foundation to develop predictive engineering tools for material processing and performance behavior and for evaluating the crash performance of plastics and composite structural panels, and other components and subsystems in vehicles for crash forces and impact geometries.

The ACC-PD conducted an informal set of interviews with automotive industry leaders, to assess their interest in PCIVs for improving the future fleet fuel efficiency and environmental performance (in the fall/winter 2006), as a complement to this effort. The ACC-PD industry survey focused on the simultaneous application of technology options in addition to light-weighting, as summarized below:

Existing Conventional Technology Options for Fuel Economy Improvement (ACC-PD 2006 Automotive Industry Abridged Interview Results)

1. VEHICLE LOAD REDUCTION
   - Aerodynamic Improvements
   - Rolling Resistance Improvements
   - Targeted Mass Reduction
   - Accessory Load Reduction

2. INTEGRATED STARTER GENERATORS

3. EFFICIENT ENGINES
- Variable Valve Control Engines
- Stoichiometric Burn Gasoline Direct Injection Engines

4. IMPROVED TRANSMISSIONS
- 5- and 6-Speed Automatic Transmissions
- 5-Speed Motorized Gear Shift Transmissions
- Optimized Shift Schedules
- Continuously Variable Transmissions

3.3 Existing Research Partnerships Relevant to PCIV Safety

The Volpe Center reviewed recent and ongoing national vehicle fuel-efficiency, safety and materials research efforts by Federal agencies and by industry and academic consortia to identify programs related to PCIV strategic research and technology deployment objectives. This revealed a number of partnering opportunities for future research and development initiatives, and for developing the requisite automotive composite materials testing standards to predict the PCIV crash performance and optimize crash energy management.

3.3.1 DOT research relevant to automotive composites and safety performance

3.3.1.1 NHTSA Research Programs

As the "Hydrogen Posture Plan"[34] prepared jointly by DOE and DOT in December 2006 indicates, DOT is a full partner in this multi-agency, industry, and academia cooperative research, development, and demonstration effort aimed to field hydrogen-fueled vehicles by 2015-2020. NHTSA is currently engaged in several research and development programs, which promise synergies with a future PCIV research and development initiative. The NHTSA Four Year (2005-2009) Plan for Hydrogen, Fuel Cell, and Alternative Fuel Vehicle Safety[35] describes several cooperative safety research initiatives in support of FreedomCAR goals. NHTSA is an active member of the Codes and Standards Technology Team, the national Hydrogen and Fuel Cell Codes and Standards Coordinating Committee, the USCAR safety Working Group and the Interagency Hydrogen and Fuel Cell Test Force. NHTSA will also monitor the real-world performance of industry prototype or developmental test vehicles (e.g., the Honda FCX, GM Equinox, and BMW dual-fuel Series 7) focusing on crash safety, fuel system integrity, and electrical system isolation. As discussed in the most recent program overview,[36] this NHTSA research effort, in partnership with the Department of Energy and industry on Hydrogen and

---

[34] *Hydrogen Posture Plan: An Integrated RD&D Plan.* (2006). Posted at www.netl.doe.gov/technologies/hydrogen-clean-fuels/refshelf/pubs.

[35] Posted at www.rita.dot.gov/agencies_and_offices/research/hydrogen_portal.

[36] See Hennessey, B.C., and Nguyen, N. T. *Status of NHTSA's Hydrogen and Fuel Cell Vehicle Safety Research Program* (ESV07 Paper No. 07-0046).

Fuel-Cell Vehicles safety already entails applications of composites to light-weight, and enable development of more fuel-efficient vehicles.

The new integrative safety approach to research and development, that aims to capture the compound safety benefits of advanced active and passive occupant protection strategies, is also relevant to the safety of future PCIVs. The NHTSA-sponsored Crash Injury Research and Engineering Network has performed studies of older driver injuries. The University of Michigan Transportation Research Institute has analyzed the injury patterns due to aging and their causes, to suggest prevention and mitigation options. The University of Michigan Program for Injury Research and Engineering Fellowship Program unites medical researchers with engineers to examine injury prevention in vehicle crashes and has focused on the elderly and mechanisms of crash injury. UMPIRE disseminates and analyzes the CIREN data, to assist the vehicle manufacturers with implementation of safety improvements based upon real life crash data. This partnership effort promises to improve the design and crash safety of future vehicles, and to better protect the elderly.

3.3.1.2   FAA Research Programs

The FAA Composites Safety Certification Initiatives[37] have focused on aerospace composites in aircraft, safety certification standards, and fire safety. Funded research and development includes composites characterization, environmental weathering, crashworthiness, non-destructive integrity evaluation methods to assess impact structural damage, damage tolerance, and maintenance practices (composite patching), as well as development of Nondestructive Test and Evaluation for composite structures. The FAA is actively partnering with standards organizations (SAE and ASTM), other Federal agencies (NASA, DOD) and industry, including the automotive manufacturers. The FAA has funded research and development on aerospace and automotive composites conducted by the Centers of Excellence university consortia. The centers are:

- The Joint Center for Advanced Materials Research (JAMS), led by the University of Washington and Wichita State University, and including Northwestern, Purdue, UCLA, Oregon State, and other partners. A JAMS effort will contribute to an updated version of Mil- Handbook-17, recently renamed the Composite Materials Handbook (CMH-17), which will consider testing procedures and the Energy Absorption data on composites important to Crash Energy Management.
- The Airworthiness Assurance consortium which includes several universities with composite materials expertise.

---

[37] Program overview presentation by Dr. Larry Ilcewicz, on May 25, 2006, to the Fort Worth DER Seminar.

3.3.1.3 FHWA Research Programs

The FHWA has ongoing research and development programs, which could be relevant to future PCIV safety, such as: application of advanced composites for infrastructure applications, non-destructive test and characterization of composites, older drivers' safety and roadway visibility enhancements for aging drivers.

3.3.1.4 FTA Research Programs

For several years, FTA has been conducting research and development programs with universities and industry partners, focusing on design, testing, and crashworthiness of a lightweight Composite Bus, Fuel Cell Buses, and Hydrogen Safety issues. Several of these programs are carried out in cooperation with the Department of Energy's Vehicle Technologies program and with State and local transit authorities.

3.3.1.5 The DOT University Transportation Centers

The TEA-21 legislation in 2000 and SAFETEA-LU legislation in 2005 authorized numerous University Transportation Centers (UTCs) and Regional UTC research consortia, to be co-funded primarily by FHWA and FTA and managed by RITA. Several UTC have recognized advanced materials expertise and focus on fuel-efficient vehicle technologies. For instance, the University of Alabama Center for Advanced Vehicle Technologies (CAVT) has been focusing on vehicle design and performance evaluations, including lightweight, composite advanced bus components and overall crash performance. The University of Missouri-Rolla strategic theme is "Advanced materials and Non-Destructive Testing Technologies," several regional UTCs, such as the Region I (lead is MIT, which has a Center for the Aging and the Sloan Automotive Center). Region III Mid-Atlantic UTC (lead is Pennsylvania State University) and Region X (TRANSNOW lead is University of Washington) have expertise in automotive materials and safety performance modeling. NHTSA could leverage a PCIV research initiative in the near-term by targeting such ongoing research and development centers and tapping into their expertise on advanced automotive materials.

3.3.1.6 Small Business Innovation Research (SBIR) Program

This set-aside DOT research and development program is managed and coordinated by the Volpe Center for all DOT modal administrations, based on their annual priorities for innovative research and technologies. The Small Business Innovation Research (SBIR) program offers the opportunity to enlist small but cutting-edge automotive materials or design companies, so as to enhance future PCIV safety.

3.3.2 Public-Private Partnership (P3) Research and Development Programs

The Department of Energy and its National Laboratories (ORNL, ANL, and PNL) are currently leading the FreedomCAR and Vehicle Technologies (FCVT) and the 21st Century Truck

Partnerships with industry.[38] These efforts build on the accomplishments of the Partnership for Next Generation Vehicles (PNGV) in the 1990s, which aimed to triple the fuel efficiency by 2000. Since 2002, the Department of Energy has led the FreedomCAR collaborative research and development initiative in partnership with industry and academia.

A primary goal of FCVT is to improve the fuel efficiency of next generation vehicles by using light-weighting materials and new power and propulsion technologies such as hybrid, fuel cells, and hydrogen fuels (FreedomFUEL). Light-weighting options for structural applications under investigation include plastics and composites, as well as metals (aluminum, magnesium, and zinc, high performance steel honeycomb structures) and their alloys.

The National Research Council (NRC) Committee on the Review of the FreedomCAR and Fuel Research Program published its first report in 2005.[39] With regard to the materials research and development program, the report concluded that the very ambitious target of 50 percent vehicle weight reduction, in order to achieve major fuel efficiency improvements at affordable cost, was not realistic. In focusing on safety as a crosscutting issue, the report identified the "critical need to develop safety-related technology, codes and standards (including vehicle standards) and inculcate widespread safety awareness." To address the hydrogen and transitional vehicles safety challenges, the report recommended that:
- The Department of Energy should form a new crosscutting safety technical team; and
- The Department of Energy and NHTSA should secure sufficient resources to carry out their respective safety roles and responsibilities.

The President's 2007 State of the Union address announced the "Twenty in Ten" fuel economy initiative, amplified by the Secretary of Transportation in her TRB 2007 address,[40] that aims to cut gasoline consumption by 20 percent in the next 10 years (by 2017) by improving vehicle fuel efficiency and through the use of flexi-fuels and transitional innovative vehicles. The PCIV may be one method to meet this goal, as well as with the Department of Energy USCAR and Hydrogen Vehicle research and development efforts.

USCAR has nine research and development consortia, but those of direct relevance to PCIV crash safety performance are the:
- U.S. Automotive Materials Partnership (USAMP);
- Automotive Composites Consortium (ACC), especially its Crash Energy Management Working Group, which conducts research on composite materials characterization, and crash performance;
- Vehicle Recycling Partnership (VRP); and
- Occupant Safety Research Partnership (OSRP).

---

[38] *Driving Technology: A Transition Strategy to Enhance Energy Efficiency.* May 26, 2006. DOE/EERE.

[39] *Review of the Research program of the FreedomCAR and Fuel Partnership: First Report* available at www.nap.edu.

[40] See the White House initiative details at www.whitehouse.gov/stateoftheunion/2007/initiatives/energy.html and Secretary Peters' speech at www.dot.gov/affairs/peters012407.htm.

The goals of the Materials Crashworthiness research and development performed by the ACC Crash Energy Management Working Group are to:[41]
- Develop and demonstrate analytical methods and design technology practices required to apply composites in crash energy management structural applications;
- Develop production representative designs, materials, and manufacturing processes;
- Develop an understanding of the physics that controls crash energy absorption; and
- Develop design methodology.

The ACC priority research and development issues for crashworthiness of lightweight vehicles are:
- To determine:
  - Where does the crash energy go?
  - What factors are important during a crash?
  - What factors are interdependent?

- To develop industry capabilities for:
  - Assembly and validation of a knowledge base of materials energy absorption behavior;
  - Methods to characterize materials;
  - Analytical models of material behavior; and
  - Validated design tools and design practices.

The Department of Energy USCAR/ACC Crash Energy Management Working Group near-term priorities are to:
- Determine experimentally the effects of material, design, environment, and loading on macroscopic crash performance to guide the design and the development of predictive tools;
- Determine the key mechanisms responsible for crash energy absorption and examine micro-structural behavior during the crash to direct the development of material models;
- Develop analytical methods for predicting energy absorption and crash behavior of components and structures;
- Conduct experiments to validate analytical tools and design practices;
- Develop and demonstrate crash design guidelines and practices; and
- Develop and support design concepts for application in demonstration projects.

The Advanced Lightweight Materials-Composite Materials 5 Years Plan by the Department of Energy/ACC partnership[42] identifies the research and development priority projects for 2005-2009, namely to:
- Identify and manufacture appropriate plastic materials that are cost-effective compared to steel and other metals;

---

[41] May 9, 2006, overview presentation by Dr. Joseph Carpenter, the Automotive Lightweighting Materials Program Manager, DOE Office of FreedomCAR and Vehicle Technologies, at the 2006 SAE Government/Industry Meeting.

[42] The ACC ALM- Composite Materials 5 Year Plan, July 2005. Provided by Dr. Chaitra Nailadi, Daimler-Chrysler Scientific Labs.

- Develop manufacturing processes that will support large volume builds (>150K/year);
- Develop reliable and low-cost joining and bonding technologies;
- Achieve high body-fit and finish (tolerances); and
- Match the Coefficient of Linear Thermal Expansion of metals versus plastics in hybrid structures.

Specific Energy Management research projects now underway focus on developing analytical tools and testing machines for crash absorption characterization of composites, sandwich, and adhesively bonded structures in axial and lateral impacts. Future projects will develop a comprehensive database of the CEM properties of automotive composites, the validation of modeling tools for automotive structures, and structural design optimization for vehicle crashworthiness. These projects, which are to be completed by 2010, promise to enable future PCIVs to comply with all applicable NHTSA crash safety requirements.

In addition to this Department of Energy led research and development program, since 2004 the Department of Energy Hydrogen Storage Grand Challenge and its National Labs have funded three university Centers of Excellence (CoE) on Chemical Hydrogen, Metal Hydride, and Carbon-based materials research and development involving over 40 universities, 15 companies, and 10 Federal laboratories. Several CoE universities have advanced automotive materials expertise relevant to the PCIV safety.

The Department of Energy National Labs play major roles in managing the USCAR partnerships and in performing major research and development programs, most closely matched to their facilities and skills base. For instance, ORNL plays a lead role in materials crashworthiness testing, PNNL in developing predictive engineering modeling software tools, and ANL in developing computational crashworthiness models for green vehicles, and in evaluating their environmental impacts and recyclability.

3.3.3 Other Federal R&D Related to PCIV

Other Federal agencies have ongoing research and development programs on advanced composites that are of potential interest to NHTSA and industry for future PCIV safety assurance. These include:
- The Department of Commerce National Institute for Standards and Technologies has just completed the Advanced Technology Program and Manufacturing Extension Program. These cost-shared research, development, test, and evaluation efforts with the automotive industry (Tier 1 OEM and plastics suppliers) have been designed to enable manufacturing of low-cost, high-quality automotive plastics and composites.
- The National Aeronautics and Space Administration (NASA) research and development on durable composites and fuel-cells for aircraft (like the composites-rich Boeing 380) and spacecraft.
- A wide range of Department of Defense research and development programs on composites applications for lightweight and more fuel-efficient military multi-modal vehicles have been conducted by various research centers, such as the USAF Wright Patterson Lab; the U.S. Army Tank Automotive Research, Development, and

Engineering Center (TARDEC) in Warren, Michigan; the Defense Advanced Research Program Agency; and the U.S. Navy/National Research Lab.

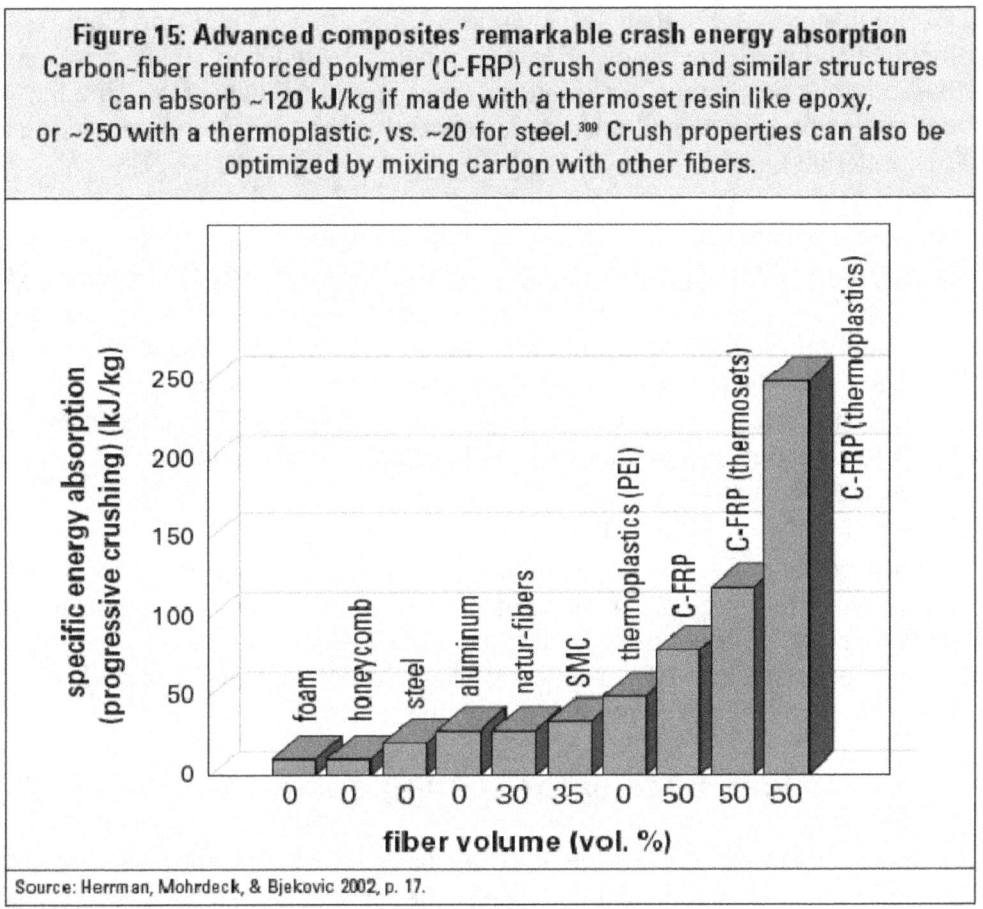

Figure 3-1 : DOE Summary of advantages for Carbon Fiber Polymer Composites in automotive and aerospace structural applications and their relatively high crash Energy Absorption (EA), when compared with alternative light weighting materials (see DOE/ USCAR references).

3.4 Standards and Guidelines for Automotive Composites Crashworthiness

3.4.1 The Society of Automotive Engineers International (SAE)

A review was carried out of the SAE standards for crashworthiness of automotive plastics under development, and of existing SAE standards for glazing materials (used in windshield and sun-roofs), and for air bags. The SAE Committee for Standardization of Tensile Testing for automotive plastics and polymers is developing standardized tests for tensile strength of plastics, and plans to develop in the next phase compressive testing standards.[43]

---

[43] Therefore, the experts interviewed in Ch. 4 included the Chair of the SAE Committee (Dr. Jackie Rehkopf, now at Exponent, formerly at Ford Research Lab); and Dr. Susan Hill, University of Dayton Research Institute (UDRI),

The SAE 2004 Auto Safety Survey Results[44] summarize Automotive Industry Safety priorities and the potential for plastics-based safety solutions. DuPont sponsored the Automotive Consulting Group to survey more than 400 attendees at the 2004 SAE World Congress on their safety development and implementation priorities. Their selected Top 10 safety systems in order of priority (based on the percentage of respondents which ranked them) from first tier suppliers and OEMs in the next five years, and related plastics and composites (without naming any specific plastic materials) are:

- Collision avoidance systems (62%) involved DuPont engineered polymers for electronic sensors packaging and new windshield materials for sharper Heads Up Displays (HUD).
- Side and rear air bag systems (56.5%) include redesigned side air curtains, knee and rear systems, including driver and occupant sensors—electronic materials include plastics for flexible circuitry, nylons, and resins for air bag deployment doors, etc.
- Active safety belt pre-tensioning systems (43.5%) use thermoplastic polyester, nylon, acetal and other resins for safety belt retractors, buckles, and motor components.
- Pedestrian protection systems (41%).
- Electronic Stability Control (ESC, 39%).
- Side door reinforcement and impact absorbers (36%) – a critical need for improved protection and vehicle-to-vehicle crash compatibility – include engineered plastics for crush cones, bladder concepts, honeycomb, and fiber materials.

Other safety items identified by SAE respondents as having lower priority include tire pressure monitors; side, overhead, and rear laminated safety glass; and inflatable seat belt systems.

3.4.2 The American Society for Testing and Materials (ASTM)

The ASTM Committee D30 on Composite materials has six technical subcommittees that maintain jurisdiction over 60 standards published in the Annual Book of ASTM Standards for testing for performance characterization of existing and emerging composites in a wide range of applications. This ASTM D30 committee is currently engaged in developing new materials testing standards for polymer matrix composites (PMC) and its activities are closely related to the ones of the Composite Materials Handbook-17 (CMH-17) organization.

CMH-17 is a collaborative effort led by the Federal Aviation Administration (FAA) and Department of Defense with the active participation of and in coordination with materials experts from Federal agencies (NASA, FAA), academia, and industry partners (aerospace, automotive manufacturers and specialty materials suppliers). Their goal is to update the three volumes Composite Materials Handbook-17 (CMH-17).

The Chair of the Crashworthiness Working Group of the CMH-17 (Prof. Paolo Feraboli, University of Washington) was one of the expert respondents to the Guided Interview and provided additional technical references (see References) and summary materials on ongoing

---

who conducted composite materials testing for the OEMs for parts used in commercial vehicles, as well as towards the SAE Standard development.

[44] Posted at www.autochannel.com/news/2004/03/03/183246.html.

PMC testing standards development and activities related to their crashworthiness. The CMH-17 process is based on roadmaps to guide the use of data and models to evaluate new materials in design and structural applications, including testing for crash safety certification of coupons, components, panels, joints, sandwiches, and hybrid structures.

3.4.3 ISO TC 61/SC 13

The International Organization for Standardization has Technical Committee, TC 61, developing international standards for *Plastics*. The TC 61 Subcommittee 13 addresses standards for polymer matrix composites and fiber reinforcements. These standards will impact the use of polymer composites in automotive structural and other applications.

3.5 International Research and Development Efforts on Automotive Light-Weighting with Composites

3.5.1 European Union Research and Development Partnerships

EUREKA is a pan-European network for market-oriented, industrial research and development created in 1985, as an intergovernmental initiative. It was designed to enhance European competitiveness through support of businesses, research centers, and universities. These consortia carry out pan-European projects to develop innovative products, processes, and services. Through a EUREKA project, the partners develop new technologies for which they share Intellectual Property Rights and facilitate the penetration of new markets.

EUREKA recently funded a program for performing and manufacturing thermoplastic composite columns for vehicles. Case study structures will be designed to comply with recent European passive safety standards and fabricated to prove the process.

The European Union has funded several multi-national "green transport" research and development initiatives, including development of light-weight, fuel efficient and low-emission vehicles, while increasing safety. The leading project on advanced, light-weight composite materials for cars is Technologies for Carbon Fiber Reinforced modular Automotive Body Structures (TECABS).[45] This four year (2000-2004) program used innovative multi-axial woven carbon fiber composites for a 50-percent body weight reduction, using 70 percent fewer parts. The TECABS partners included auto manufacturers (Volvo, Renault), materials suppliers, and five leading universities. Other advanced composite projects were the Fiat-led "3D-Structures" and the HYDROSHEET project, aimed at reducing the weight while increasing strength and crash safety through the development of rapid manufacturing and resin curing processes. Future commercial vehicles developed as a result of these international public-private partnerships would have to meet the U.S. crash safety standards, as well as EPA emission regulatory limits.

Europe (Bayer) and Japan (e.g., Fujitsu, Toyota Motor Co.) also led in the development and application of bio-based polymers, which are derived from plants (rather than petroleum), and

---

[45] See postings at www.tecabs.org and news items on the European transport research.

are therefore biodegradable and recyclable. Toyota has used bio-plastics since 2003 in its Prius and Raum vehicles, for both interior and structural applications. Toyota expects that about 20 percent of the world's plastics will be biodegradable by 2020, and has become a producer and supplier of bio-plastics from renewable resources.

Although plastics use a small percentage of petroleum and natural gas, there is a perception that the availability of plastics may be limited in the near future. Almost any hydrocarbon source can be used for plastic and plastic composite feed stocks and research is expanding relative to the future increased use of coal or biomass. It is likely that alternative, bio-based sources of plastics and composites will eventually be developed, at low cost and high volume, thereby enabling large scale deployment of sustainable PCIVs. The crash and other safety performance of emerging biopolymers would have to be evaluated (e.g., fiber-based materials absorb moisture and expand).

U.S. research on the production of biopolymers (from corn, sugar cane, and tapioca) has been performed by government, industry, and academia[46] (e.g., the Department of Agriculture, Cargill/NatureWorks, ADM/Metabolix), including automakers (Ford, John Deere). However, the cost of bio-plastics is currently higher than that of petroleum-derived conventional plastics, and their structural strength, durability, and other properties (e.g., water absorption, thermal expansion) would have to be much improved before integration into future PCIVs.

Our analysis indicated that there are major challenges to overcome for successful development and fielding of PCIVs by or in 2020, which are not related to safety. Since plastics are currently derived from petroleum, their cost and availability would be affected by rising costs. Furthermore, since energy independence is a major national policy goal, alternative sources of plastics are desirable. Any substitute material to oil-derived plastics and composites in automotive structural applications will have to meet crash-performance specifications.

3.5.2 Research and Development Partnerships in Japan

The "CFRP Automobile Project" in Japan was described at the recent United States-Japan Conference on Composite Materials (September 2006, in Dearborn, Michigan).[47] This is a five-year (2003-2008) partnership effort of Toray Industries and Nissan, cosponsored by the Ministry of Economy, Trade, and Industry (METI) as the NEDO program. Its objective is to design a Body in White (BIW) with half the weight and 1.5 times higher crash energy absorption than a steel-body conventional car, using continuous fiber fabric reinforced thermoset resin materials. The program is also addressing manufacturability, processing time, recyclability, and cost issues, to ensure that a crashworthy BIW concept is developed and deployed.

---

[46] See publications on biomaterials for structural applications by Prof. Drzal and colleagues at the Michigan State University Composite materials and Structures Center at www.chems.msu.edu.

[47] Kitano, A., Wadahara, E., and Taketa, I. *The CFRP Automobile Project in Japan*. Paper presented at the 12th United States-Japan Conference on Composite Materials, September 21-22, 2006. Dearborn, Michigan: University of Michigan.

3.6 Current Trends, Best Practices, and Lessons Learned For Automotive Composites Integration

3.6.1 Emerging PCIV Concepts and Best Safety Practices (BSP)

With the increasing aging populations in the United States, Europe, and Asia combined with new emerging power plants (hydrogen internal combustion, fuel cell stacks, electrics, diesel-electric hybrids and others) and the need to enhance fuel economy and reduce greenhouse gas emissions, future vehicles will incorporate lightweight materials, facilitating PCIVs by 2020. State initiatives have been introduced in California and the northeast to improve fuel efficiency of the vehicle fleet, and to reduce the transportation environmental impacts. For example, Montana created a new vehicle category called Medium Speed Electric Vehicles (MSEVs). MSEVs, like the Zenn electric car and Chrysler neighborhood vehicles, are plastic intensive lightweight vehicles being sold in the United States today. They can be driven at speeds up to 35 mph, and represent a niche market that could speed up commercialization of innovative vehicles. Washington and California have similar legislation pending.

Innovative lightweight concept cars that make use of advanced plastics and safety features have often been shown at auto shows, but the challenge remains bringing them into production for mass markets. Designing the future PCIV for superior safety performance can build on industry Best Safety Practices and on lessons learned over the past decade, such as the successful designs and superior performance of high-end sports and racing cars, and concept cars that used advanced composites higher speed performance, and enhanced crash safety (e.g. Ferrari Enzo, Porsche Carrera GT). For instance, the 1998 Chrysler Composite Car (CCC) was designed as a five-passenger lightweight, fuel-efficient vehicle (50 mpg), from fiberglass composites with 15-percent plastics content derived from recycled plastic bottles. However, it was designed for the emerging (China) market and could not meet the U.S. crash testing requirements.

Minicars that can meet the current crash safety standards are already being sold in the United States. They incorporate to varying degrees plastics, plastic composites, and plastic hybrid materials for interior and structural applications. BMW's Mini Cooper and DaimlerChrysler's SMART car have had market success in Asia. The Mini Cooper has been already sold in the United States, and the SMART Fortwo and Forfour will be marketed in 2008. The subcompact Chrysler Smart Fortwo vehicle features a solid steel frame for crash strength and a front crumple zone, with interchangeable, light, and recyclable plastic body panels. Similar vehicles are under consideration by automobile manufacturers in response to higher fuel prices, a possible change in U.S. fuel-economy regulations, and changes in Americans' buying patterns for small vehicles to mitigate climate change and improve fleet sustainability.

Plastics and plastic composites are gaining popularity with automobile manufacturers for the compact car segment. Several polymer intensive vehicles are slated for introduction within the next few years. Industry leaders have recently developed concept vehicle designs that use advanced materials for both fuel efficiency and for crash safety and promise early deployment. Among the concepts shown in 2007 in Detroit, New York, and Los Angeles, and at international auto shows:

- The Ford Edge, with HySeries Drive technology, was demonstrated as the first plug-in hybrid vehicle with a hydrogen fuel-cell, although it features a lightweight aluminum body, rather than plastics.
- General Motors introduced its new minicar platform, designed to serve a variety of distinctive body styles, and to meet crash safety standards in the United States, Europe, and Japan. The Chevrolet Beat, Trax, and Groove are three minicar models which will share the GM global platform architecture. A common feature is the use of plastics for exterior body panels, as well as extensive interior and structural plastic, plastic composite and plastic hybrid materials integration. GM stated that one of the three concepts will be selected for production by 2009, to be sold in the United States and worldwide.
- The GM concept car Chevy Volt is a new plug-in series hybrid, which is lightweight and compact.[48] It features a new electric propulsion and flexi-fueled (E-Flex) system. Lightweight components use resins and composite materials developed by GE Plastics to save 30-50 percent of weight, yet offer superior strength for the frontal and rear energy absorbers (e.g., high-performance thermoplastic composite [HPPC] in doors, hood, and front fenders) and polycarbonate glazing.
- The Hyundai concept QarmaQ, shown at the 2007 Geneva Auto Show, was developed as a joint project with GE Plastics and uses a wide range of plastic composites. Although it is only 60 kg lighter than if it were made from conventional steel (using about 900 recycled plastic bottles per car), its "elastic front" plastic nose design promises improved crash safety for both occupants and pedestrians, thanks to better energy absorption and flexibility.

Other recently advanced fuel-efficient vehicle design concepts were developed by non-profit or public interest organizations to demonstrate the fuel efficiency achievable with advanced materials and current technologies, but not commercialized as yet. They utilize lightweight materials and were designed for equal or improved crash safety:

- The Rocky Mountain Institute and Hypercar, Inc. designed in 2000 the Revolution concept "hyperlight hypercar."[49] It is a mid-size five-passenger SUV crossover with half the weight of a conventional analog, which achieves superior crashworthiness by using carbon-fiber composites and aluminum for the structural strength of both the body-in-white frame and the passengers "composite safety cell." This advanced fuel cell vehicle (FCV) was also designed with an integrated digital vehicle dynamics control and suspension, for a novel manufacturing and assembly process, to cite: "…new design and manufacturing opportunities can make advanced composites the best choice for replacing steel, to save over 60 percent of BIW mass."
- The Union of Concerned Scientists (UCS) developed the Vanguard minivan blueprint, designed to meet the new California GHG emission standards while maintaining equivalent levels of safety and performance.[50]

---

[48] See *GE's weight reducing contributions to the Chevrolet Volt*, January 14, 2007. Posted at www.greencarcongress.com/2007/01/ges_weightreduc.html.

[49] Lovins and Cramer. (2004). Hypercars, Hydrogen and the Automotive Transitions. *International Journal of Vehicle Design, Vol. 35, Nos. 1-2*, 50-85.

[50] See postings at www.ucsusa.org/clean_vehicles for details on the Vanguard and other proposed and existing fuel efficient vehicles (40 mpg fleet is considered feasible within 10 years with today's technologies).

Barriers to the increased utilization of advanced composites in commercial vehicles include their relatively high cost (both materials and retooling the manufacturing and assembly process), as well as the still unproven lifecycle cost-effectiveness, durability, reliability and maintainability. Even if the knowledge gaps and R&T priorities identified below in Chapter 4 were to be successfully met by 2020, these attributes remain the key concerns for the successful commercial PCIV deployment in the future. Best practices can help overcome these barriers. Below are listed examples of industry "best practices" and case studies for increased use of plastics and composites in commercial vehicle applications.[51]

Partnerships: Partnering between composites materials suppliers, designers, and modelers and vehicle manufacturers is a proven best practice. The pre-competitive and cost-shared cooperative research and technology undertaken by public-private partnerships (like USCAR) and university consortia is another BSP relevant to future PCIV safety. Accomplishments can subsequently be commercialized by the participating industry partners in their various vehicle models.

In the PNGV, program successful commercial partnerships for composites (cited by Brylawski and Lovins) included the Bayer/GE Plastics development of polycarbonate glazing and of DSM/BASF for structural resins, etc. Trade associations, like the American Chemistry Council-Plastics Division could facilitate industry inter-sector partnerships. In the PNGV program, this was accomplished by the steel industry for the Ultra Light Steel Auto-Body (ULSAB) and by the aluminum industry, to develop lightweight prototype vehicles and associated design concepts and manufacturing processes.

Standardization and global platforms: A promising strategy to overcome the barriers identified below in Section 4.2.5 by the Automotive Composite Consortium (ACC) might be to use "standardization" and "global platforms."[52] For instance, common and modular hybrid drive-trains were developed, standardized, and integrated into a broad range of vehicles to reduce costs and enable scale-up, as done previously for Antilock Brake Systems (ABS), to enable widespread adoption. "Global platforms" (adopted by Ford, Toyota, Hyundai, Renault-Nissan, VW, and others) use common building blocks and vehicle architectures (basic body materials and geometries, mounting points, standardized manufacturing processes) for families of vehicles, to cut cost and allow high-volume production of subsystems.

Voluntary measures to improve safety: Industry leaders in automotive safety have developed and deployed advanced safety systems on a voluntary basis and derived business benefits from publicizing them. Examples include Volvo's Whiplash Protection System designed to prevent and mitigate head-neck injuries in crashes, the Honda Advanced Compatibility Engineering (ACE) structural designs, etc. These business best safety practices are expected to accelerate and expand for next-generation vehicles, since "safety sells" and the public expects higher levels of safety and reliability.

---

[51] Hazen and Musselman in the 2006 *Automotive Composites-Design and Manufacturing Guide.*

[52] See articles: The Power of Platforms in *Automotive Engineering* online, January 2007; and Developing Hybrid Drive Systems for a Broad Range of Vehicles, ibid.

Voluntary measures that address compatibility challenges are relevant to the safety of future light-weighted PCIVs. The automotive industry has voluntarily adopted improved crashworthiness and compatibility vehicle designs, and developed a wide range of fuel-efficient concept cars which appear to hold promise for safety as well. These include improved structural alignment implementation of Secondary Energy Absorbing Structures (SEAS) and the Honda Advanced Compatibility Engineering (ACE) body structure to minimize the impacts of a mismatch in size and weight. Integrative safety solutions that are being explored include the benefits of combined collision avoidance prevention strategies with existing and emerging active and passive safety crash impact mitigation systems. As an example, in 2003 the Alliance of Automobile Manufacturers adopted voluntary crash compatibility enhancements like the Electronic Stability Control (ESC), to be completed by 2009 in advance of the mandated 2012 phase-in date.

Another recent example is Ford's rollout of its advanced safety canopy, side air curtains, and rollover sensor as standard equipment on fourteen 2007 vehicle models, and promise for early introduction of rear seat inflatable seat belts shown on the Ford Interceptor concept in 2007. The advanced four-point seat belt technology evaluated in Mustang Cobra tests for user friendliness, and shown in the Ford Interceptor and Airstream concepts promises to improve safety for older drivers and passengers.

Peer recognition: The best automotive plastics applications are evaluated annually by the Society of Plastics Engineers (SPE) Automotive Division by a panel of peers and recognized with awards, including a safety category. This type of peer recognition rewards the teaming of plastics suppliers, engineering design firms and the OEMs in developing and commercializing new designs or advanced materials applications, and the integration of innovative assemblies. Currently, there is no SPE award for "weight reduction for fuel-efficiency with plastics," because each and every nominated plastic component is reviewed by the judges for weight reduction, cost reduction, and parts elimination, and thus would not warrant a dedicated category.

At the World Traffic Safety Symposium, DOT and NHTSA representatives served on a panel of judges along with other industry and association safety stakeholders, to recognize innovative automotive safety features. The 2007 award recognized Volvo's improved child protection integrated booster seat system, combined with inflatable curtains and load-limiting seat belts. The X-Prize Foundation has announced in 2007 an Automotive X-Prize (AXP)[53] competition offering a multi-million dollar award for a super-efficient 100 mpg (or equivalent), production-ready automobile.

3.6.2 Lessons Learned

Useful "Lessons Learned" for future PCIV safety performance can be derived from the utilization of advanced plastics and composites in structural and safety applications for racing cars (Formula-1 and NASCAR), concept cars (Chrysler Composite Car[54] [CCC], VW EcoRacer,

---

[53] See http://www.xprize.org/xprizes/automotive_x_prize.html

[54] In general, this report refrains from mentioning specific materials or industry suppliers, in order to avoid perceived favoritism. Any industry or trade product names appear merely for illustrative purposes, and should not be construed as endorsements.

FORD Reflex, GM Hywire), and high-end sports supercars ("roadsters"). Their designs feature lighter and tougher aerospace materials as panels or monocoque bodies (carbon fiber reinforced composites), and demonstrate superior crash safety performance, both in impact resistance and occupant survivability. The advanced composite materials used in motorsports to reduce subsystems and body weight, while enhancing their crash safety, were discussed at the 2007 Global Motorsports Congress in Germany.

Other light and strong materials are used in the space-frame, wheels, and other structural and non-structural elements of sports vehicles, such as metals (aluminum, magnesium, titanium, and high-performance steel). Furthermore, the goals in using these advanced materials in sports cars are to achieve high speeds and high acceleration at equal safety, not to improve fuel efficiency.

Experience gained to date, however, may be applicable to future PCIVs, although these categories of vehicles are typically produced at very low-volume and high-cost. NASCAR racing cars (like Formula-1) and high-end, high-performance sports vehicles (e.g., Mustang, Corvette, and Viper), which make extensive use of plastics and composites in their monocoque shell body, wheels, and other chassis and propulsion subsystems, have also met NHTSA crashworthiness requirements. Their superior safety performance and occupant protection strategies for high-speed impact survivability include: four point belt restraints, Head and Neck Support (HANS) system and composite helmets, and nonflammable body suits. If modified, clearly some, but not all, of these might be transferable to commercial and personal vehicles, assuming that the high costs of materials and manufacturing can be reduced by economies of scale at high-volume mass-production.

Another lesson can be learned from the use of plastics in commercial cars, specifically the plastic body and door panels for GM Saturn cars, which have satisfied all NHTSA safety standards for over 16 years. Based on trade news articles, GM discontinued the Saturn plastic body in 05-06 models because of fit and surface finish issues. Issues included gaps between panels in cold weather due to mismatched linear coefficient of thermal expansion (CTE) of thermoplastic materials used, which also contributed to door expansion in hot weather, and to difficulty in extricating injured or trapped passengers after a crash.

The literature claims that U.S. vehicle manufacturers have developed and marketed in Europe smaller, lighter, and more fuel-efficient vehicles, which are not available in the U.S. Generally, European and Japanese cars are smaller, lighter, and more fuel-efficient than U.S. counterparts, and make greater use of plastics at equal or better safety performance.[55] Further study is needed to provide concrete evidence for this and to evaluate transferability as a best practice to the U.S. market and highway environment. In Europe, the automotive plastics are also fully recyclable to comply with the End-of-Life Vehicle (ELV) requirement. Similarly, U.S. manufacturers tend to use lower amounts and different types of plastics in their U.S. models, although they incorporate more plastics in their lighter car models distributed in foreign markets. There is a need to examine and compare the relative crashworthiness of similar models and size vehicles, in order to identify and overcome the barriers for U.S. manufacture of future PCIV designs.

---

[55]Recent trade news including: Can US adopt Europe's Fuel-Efficient Cars? (2007, June 26). *Wall Street Journal*,; and *Top 10 Fuel-Efficient Cars*; *Which are the Most Fuel-Efficient Cars?*; *Why aren't US-based car makers already building great small cars?*; and *Small cars to be big part of Ford comeback.*" Retrieved from www.About.com/Cars

Several authors of references listed in Chapter 7 (e.g., Lynn, Lovins, and Cirincione) discussed the lessons learned from the Public-Private Partnership for a New Generation of Vehicles (PNGV) in the early through mid-1990s, which aimed to triple fuel-economy by 2000 through mass reduction, use of advanced materials, and other innovations in power and propulsion technologies. For instance, in the late 90s the Advanced Composites Consortium designed and built a Ford Escort glass-composite front end which was 25-percent lighter than the steel counterpart, yet passed all Federal crash safety tests with superior performance even without air bags (cited by Brylawski and Lovins, 2000).

The PNGV materials options studied and demonstrated for body and chassis construction with structural strength at reduced weight included: aluminum unibody, aluminum and steel space-frames with polymer composite panels, and polymer composite monocoque, or steel honeycomb structures. These innovative designs and technologies could be applied to PCIVs if materials processing complexity and duration, quality control and manufacturing costs could be substantially reduced for high volume vehicle production. Although much technological progress was made under the PNGV program, its goals were not fully accomplished and prototypes were not commercialized.

4. SURVEY OF EXPERTS AND SUMMARY OF FINDINGS

4.1 The Experts' Survey Design

One method for using experts' opinions to assist in decision-making under uncertainty is the so-called Delphi Approach. In order to complement the literature review summarized in Chapter 3, a focused Delphi survey of leading experts on automotive plastics in structural or occupant safety applications was designed and conducted.

A structured survey interview guide was developed (Appendix 4.1) summarizing the key issues of interest to a future research initiative on PCIV safety, and to a safety technology integration roadmap for automotive plastics and composites. Specific recommendations were sought on PCIV safety-related applications, including:
- the technology deployment process to-date versus the future projected timetable for integrating new materials and safety devices into commercial vehicles;
- knowledge gaps and research priorities for automotive plastics in the near-, mid- and long-term;
- the research opportunities with highest leverage for a potential NHTSA research partnership initiative on PCIV safety; and
- the barriers that would have to be overcome to deploy advanced plastics and composites in safe and lightweight PCIVs by 2020.

The pool of recognized experts was identified with APC, NHTSA, and Department of Energy assistance, as well as based on the technical literature review. They were first contacted by phone or e-mail to ascertain their interest and availability for brief interviews. The experts on automotive plastics and their value in enhancing future PCIV safety performance were broadly representative of all stakeholders and included:

- NHTSA experts familiar with crash safety performance and occupant protection, and/or with crash data for aging drivers and passengers (from the Office of Vehicle Safety Research, the Vehicle Research and Test Center (VRTC), and the Crash Injury Research and Engineering Network (CIREN);
- Department of Energy experts (both from Headquarters and from National Laboratories), who manage cooperative research efforts on composites crashworthiness modeling and testing;
- Tier 1 plastic materials suppliers, who understand automotive safety needs and solutions for Original Equipment Manufacturers specifications for design, materials, and crash performance;
- Automotive industry experts (from Ford, GM, and DaimlerChrysler), who understand the practical aspects of plastic components integration process, from research through deployment into commercial vehicles, and who participate in the USCAR Automotive Composites Consortium Crashworthiness Working Group;

- Members of professional and technical associations and their standards committees familiar with both plastics and automotive safety: Society of Automotive Engineers Cooperative Research Program and Standards Development Committee, and the Automotive Division of the Society of Plastics Engineers;
- Representatives of non-profit stakeholder organizations (e.g., the Insurance Institute for Highway Safety); and
- Leading university researchers, who either specialize in safety issues of older drivers, or develop and test the safety performance of aerospace or automotive plastics.

The survey was conducted via phone interviews and e-mail exchanges. About half of the experts contacted responded (see Appendix 4.2). In order to comply with OMB guidelines on surveys, the brief telephone interviews conducted were tailored to the specific research or development expertise of each respondent, and several also completed the written Interview Guide.

The subject matter experts were asked by the Volpe Center team to:

- Select their recommended "Top 3" specific safety applications of composites with greatest safety benefit;
- Identify the specific Plastics/Composites materials and technologies for each safety application;
- Indicate how and to what realistic extent they could/would better protect aging drivers and passengers, if implemented by 2020; and
- Suggest how to test and demonstrate quantitatively safety improvements for the elderly in earlier (research, development, test, and evaluation or prototyping) stages.

The inputs from the structured expert interviews facilitated the identification of consensus research and development priority needs in the near-, mid-, and long-term summarized below. The experts provided valuable inputs regarding knowledge gaps and barriers to greater deployment of automotive plastics, as summarized below.

4.2 Summary of Experts' Inputs on PCIV Safety Priorities

4.2.1 Knowledge Gaps in Predicting the Crash Performance of Plastics and Composites:

- Develop and Standardize Test Protocols for composite materials over broad range of strain rates (specimen sizes, shapes, and stress geometry and rates) to derive failure thresholds and input parameters for predictive mechanical models.

- Understand the materials-specific Energy Absorption for parts configured to improve energy absorption throughout the vehicle structure.

- Standardize engineering classification and naming of composites based on physical/mechanical properties (as done for steel grades).

- Standardize fatigue tests for aged plastics/composites.

- Identify all possible damage and failure modes as function of material structure, manufacturing process, bonding, and safety application.

- Improve predictive engineering models for crashworthiness of composite materials to consider:
  o Fundamental material crush properties of structural plastics/composites;
  o Yield and strain to failure at high strain rates;
  o Creep, fatigue and aging, and behavior under extreme environmental conditions (humidity and temperature) over the lifespan of the vehicle (>10 years);
  o All damage and failure modes for composite materials, with arbitrary structures under arbitrary loads;
  o Energy Absorption dependence on size, geometry, and impact force configuration; age and environmental effects processing; manufacturing; and design factors;
  o Crush initiators and high strain rate response for crack propagation prediction; and
  o Performance of dissimilar materials in hybrid structures.

- Standardize PCIV designs and composites for safety applications so that the safety advantages can be well proven, and the costs can be reduced for high-volume production.

The interviews with experts were conducted by phone, with either individuals and/or teams. Notes were supplemented by the completed Interview Guides and supporting published papers, which were attached to e-mail replies (as indicated in Appendix 4.2.).

Selected Quotes from Experts Interviews:
- "Multiple materials solutions are available to solve any safety problem…plastics are no better or worse."
- "There is no basic understanding to predict the performance of automotive structural composites in real crashes."
- "There is no correlation between testing of small composite specimens and the crash failure behavior of a full-size vehicle in real crashes."
- "Safety standards are written for metal—not composite—vehicles."
- "Laminates and sandwiches are expensive materials and components… must determine their damage tolerance within three to five years before going into production…we are not ready to produce PCIVs now, can't use the existing knowledge base."
- "Industry can now create vehicles that are equally safe, using all materials candidates… but the current advantage is weight loss, not enhanced safety! They can design vehicles to meet all safety criteria, within the cost and design (volume, packaging) constraints."
- "Other than using aerospace composites in structural applications in high-end cars, there are NO clear safety advantages, there are several equally safe materials to use, at lower cost. (Some stronger glazing materials are under development.)"
- "To realize the x3-5 better energy absorption in safer designs for CFC structures, we need large scale deployment."
- "If Congress passed a law tomorrow requiring PCIVs for niche markets, we still can't do it: there is not enough CFRC material, and a high cost to implement production, even if

the cost of fiber polymers were low and the supply abundant: tooling, bonding technologies, manufacturing, assembly, certification are all barriers."
- "The current restraints system protects everyone, not just the elderly."
- "The issue is not so much the technical challenges, but is more about the lack of design and integration of advanced composites into current vehicle development. If a major OEM released an advanced composite component, the market would respond."
- "From a safety standpoint, energy absorption and dissipation is key and is why metals have been the preferred choice of materials to manage energy during crash."
- "Regulations for PCIV should be the same as for current passenger vehicles."
- "There should not be any reduction in the performance requirements regardless of the materials used in the vehicle. Materials must meet current performance requirements".
- "There should not be any compromise in safety to facilitate the increased use of any material."

4.2.2 Research Needs to Predict the Crashworthiness of Automotive Composites

The high-priority research[56] needs that would enable greater utilization of automotive plastics and composite materials in future PCIVs include the:
- Development of full three-dimensional analysis modeling tools;
- Understanding of how failure and energy absorption are controlled by processes at several length-scales;
- Inclusion of all damage modes (and associated failure models/criteria);
- Consideration of interaction effects in crashes;
- Standardized tests for fatigue, creep, and aging effects;
- Consideration of structural configurations in impact crash performance;
- Understanding the issues related to manufacturing and lifetime handling;
- Consideration of system-wide effects in crashes;
- Inclusion of probabilistic failure aspects; and
- Identification and proper modeling of the actual crash reality (geometry and force).

4.2.3 The American Chemistry Council - Plastics Division Survey

An independent and complementary industry survey was conducted by the ACC-PD principals.[57] In 10 personal interviews with vehicle manufacturers and Tier 1 suppliers, it targeted the higher-rank officials (Vice Chairmen, Vice Presidents, product engineers, regulatory managers, and directors of vehicle platforms) responsible for vehicle design and safety performance. The ACC-PD summary of industry interviews[58] noted that there is industry interest in plastics for light-

---

[56] These knowledge gaps were identified by Prof. P. Lagace, MIT for the CEM-WG in October 2006, and complemented by inputs from Prof. P. Feraboli, the CEM-17 Chair and Dr. J. Rehkopf (SAE Composite Testing Standards Committee Chair).

[57] Dr. M. Fisher, Mr. James Kolb, and Ms. S. Cole also served as a peer advisory group for the entire study.

[58] Cole and Associates, Inc. (December 2006). *Industry Interview Summary - ACC/APC Enhancing Automotive Safety with Plastics.*

weighting and for safety enhancements, especially for protecting aging drivers, but that cost, manufacturability, regulatory, and other challenges remain. There was interest in greater use of plastics and plastic-metal hybrids for improved safety performance of:
- the hood as the "head impact zone for pedestrian collisions;"
- bumpers for lower leg impact zone and pedestrian protection;
- "breakaway" softer parts for vehicle interiors;
- plastic foams for front header, side air bags, and head impact countermeasures;
- adaptive restraint systems and force limiters for aging drivers;
- balanced weight reduction of front and rear seat assemblies and lowering the center of gravity to prevent rollovers; and
- structural applications.

4.2.4 Priority Research Opportunities for Future PCIV Safety

Subject Matter Experts who were interviewed identified the following priority research topics by time horizon:

A.   Near-Term Research and Development Priorities (three to five years) include:

- Stronger foam filling on side doors and posts, combined with soft foams to mitigate side-impact intrusions;
- Squirted, rigid "structural foams" to fill in and reinforce metal roof structure and pillars mitigate rollover injuries;
- Use of plastics in roofs to lighten top-heavy vehicles and lower center of gravity;
- Lighter but stronger vehicles (e.g., replace metal pillars with tubes filled with structural foams);
- Improvement of cushioning and belt restraints (e.g., use woven cylindrical seat belts, four-point attachments);
- Adaptive restraint systems "tuned" to occupant size, weight, and age;
- Use of "smart" materials for "smart" safety devices; and
- Standardization to the best and safest subsystems designs (such as head restraints, seat system designs, etc.).

A cross-functional industry team identified additional near-term research objectives, to address specific NHTSA safety requirements in the relevant Federal Motor Vehicle Safety Standards (FMVSS) by using:
- Interior plastics and foams to address applicable NHTSA safety requirements (FMVSS nos. 208, 214, 201 and 201U, 207);
- Body enhancement foams that address NHTSA regulations (FMVSS nos. 208, 214, 216);
- Seatbacks responsive to standards (ECE17, FMVSS no. 202A); and
- Exteriors structural strength – per FMVSS no. 215 Exterior Protection in CFR Part 581.

B.   Mid-Term Research Priorities (5 to 10 years) include:

- Computer-based models for crashworthiness prediction;

- Validated composite components;
- OEM design guidelines for automotive composites;
- Validated crashworthiness performance of Carbon Fiber Reinforced Composites using improved:
  - Testing standards for high-rate impacts;
  - Energy absorption predictive tools;
  - 3-D computer modeling of materials behavior versus time;
  - Durability testing standards;
  - Verification in full-scale field testing; and
  - Integrated designs for active belt, air bags, and seat systems to enhance protection in side impacts.
- Development of new PCIV designs (three to seven years);
- Marketing of successful PCIV prototype (7 to 10 years); and
- Industry-identified priorities include:
  - New Federal requirements (FMVSS) for vehicle-to-vehicle compatibility development that appropriately accommodate PCIVs; and
  - Interior and exterior plastic applications and body engineered systems to support new FMVSS requirements.

C. Long-term Research Priorities (10-15 years) include:

- Make use of improved fiber reinforced plastics for rigid door panels, to tailor energy absorption to depth of deformation in side crashes;
- Improve vehicle-to-vehicle crash compatibility;
- Lightweight the entire fleet, or lightweight the heaviest vehicles;
- Compensate with improved passive and active safety devices any disadvantage of lighter weight and smaller size cars in collisions with larger and heavier vehicles; and
- Use advanced materials in automotive safety applications, such as nano-composites, hybrid polymers, bio-polymers, and natural fiber materials (which are claimed to be eco-friendly and recyclable to meet End-of-Life goals), but only if they can be shown to perform optimally in crash situations.

The industry-identified priorities also include the development of high-performance polymers that meet both cost and safety needs.

4.2.5 Research Needs for Occupant Safety

- Improve statistical crash data analysis to understand how severity of injuries and survivability vary with age, and identify mitigation options.
- Develop stronger passenger compartment designs with frontal crush boxes.
- Improve the occupant restraints and seating systems to restrict side head movements and limit head and neck injuries.
- Adaptive restraint systems "tuned" to occupant size, weight, and age or fragility (to "normalize" injury criteria).

- Reduce impact loads with customized occupant space (seating, bolsters, belt system) for improved protection and comfort.
- Optimize the design and performance of the combined passive and active restraints system ("sum total of interior passive foams, active air bags and belts").
- Other industry-identified priority PCIV safety applications include:
  - Four-point seat belts and seat belt limiters that can do more to protect aging drivers than a body structure that meets FMVSS requirements;
  - Plastics that have strain-to-fail characteristics similar to steel that are not strain rate or temperature sensitive;
  - Vehicle structure that produces a similar vehicle crash pulse as current production vehicle structures using metal (aluminum or steel);
  - Enhanced visibility (glass composites to reduce nighttime glare); and
  - Pre-crash sensors for gentler deployment of safety devices (smart air bags, load limiters, inflatable seat belts).

4.2.6 Barriers and challenges to PCIV development and deployment

The challenges to market adoption of composites, prioritized by the USCAR Automotive Composites Consortium Board:[59]
1. Total Accounted Cost
2. Manufacturing (Feasibility and Variability)
3. Body Construction (Bonding, Joining and Assembly, Feasibility and Variability)
4. Tier 1 Supplier Capability
5. Crash (Energy Absorption and Structural Integrity)
6. Material Performance
7. Durability
8. Recycling
9. Design Methods and Capability
10. Electromagnetic Compatibility (EMC) to Radio Frequency Interference (RFI)
11. Reparability and Maintainability
12. Flammability
13. NVH Performance

Other barriers and challenges identified by the Volpe survey and literature review are:
- The high cost and limited supply of carbon fiber;
- The high cost to implement vehicle production (for retooling, bonding and processing technologies, manufacturing, assembly, safety certification), even if the cost of fiber polymers were low and the supply abundant;
- Manufacturability and production retooling;
- Rapid and reliable joining methods;
- The need to certify crash performance for new materials and/or new applications; and
- The current, complex, and long OEM process to select, test, integrate, and field composites in commercial vehicles. Below are typical durations of each step:

---

[59] Written communication to complement telephone interview of the ACC-Crash Energy Management (CEM) Working Group (WG)

- o Screen and select materials for performance and cost: one to two years.
- o Develop and evaluate design (modeling): 6 to 12 months.
- o Develop and implement manufacturing process: three to six months.
- o Perform verify and validate: 6 to 12 months.
- o Field test prototype (crashworthiness, durability, creep, fatigue, aging): one to two years.
- o Integrate technology into production vehicles: two to five years.
- o Pass NHTSA safety certification and voluntary standards (NCAP, SAE): one year.

4.2.7 Suggested Strategies to Overcome Barriers to PCIV Safety Deployment

The experts suggested several strategies to speed up the integration of novel advanced materials and their safety applications in the next generation of commercial vehicles. Selected representative citations are provided below for future consideration:
- "The only way to qualitatively and quantitatively prove new materials/applications for all drivers is to improve anthropomorphic dummy design and feedback as well as increasing the range of seating positions beyond 5% and 95% percentile dummies."
- "There will always be some level of testing that is required to validate and quantify the performance of the component, system and vehicle during crash."
- "Computer Assisted Engineering and Design (CAE/CAD) tools can be used prior to validation testing to quantify performance, but the output from the CAE tools is only as good as the material data that is used as input for the simulations."
- "It is critical to be able to characterize the materials under the same operating conditions as seen in the field. The first step is to characterize the materials and use CAE to develop a material model. Second step is to develop a lab scale validation test so that material models can be verified to CAE. Third step is full vehicle CAE models using validated material models. Last step is full vehicle crash test. These results can be compared to the full vehicle CAE crash model."

4.2.8 Suggested NHTSA Role and Opportunities for PCIV Safety R&D

The experts consulted were enthusiastic about potential NHTSA participation in cooperative research efforts on light-weighting and standards development activities. They made the following constructive suggestions for consideration towards a future NHTSA PCIV initiative:
- Provide industry-wide forums to exchange ideas and results of pre-competitive research;
- Provide funding for Working Groups and organizations (SAE) engaged in pre-competitive research;
- Participate in development of guidelines and standards for crashworthiness of structural composites;
- Establish and/or certify databases on materials properties for industry use;
- Be an integral part of current ongoing research early in the development phase;
- Crash test, analyze, and monitor performance of new PCIV prototypes in partnership with industry;

- Participate in the design, or monitor industry standardized crash testing, of prototype PCIV vehicles to demonstrate any safety advantages of strong, high-energy absorbing composites versus alternative materials;
- Require adaptive safety appliances to enhance protection of older occupants (e.g., knee bolsters, side air bags, and curtains);
- Optimize combined passive and active safety system so as to protect the older and most fragile, rather the average adults; and
- Exploit additional passive safety opportunities (mandate side curtain air bags, knee bolsters, etc.)

4.3 Recommended Top-3 PCIV Safety Research and Development and Technology Integration Priorities for Roadmap Development

Several research topics suggested by the experts also appear as priority activities identified by the ACC-PD workshop (Appendix 3.1). Those selected for the Safety Roadmap development in Chapter 4 have near-term aspects (e.g., development of improved predictive tools and certified databases on the mechanical properties of advanced automotive composites) that can be continued in the mid-term, such as verification and validation of the improved crashworthiness modeling tools developed.

Similarly, the most promising mid-term activities should also have promise and payoffs for long-term PCIV safety technology integration and deployment. For instance, PCIV prototyping and crash testing are needed to demonstrate enhanced protection for all occupants, including the elderly. This is the basis for recommending the Top 3 priority activities in each temporal category below:

4.3.1 Near-term Priorities (3-5 years) for Research, Development, and Technology

1. Develop improved statistical crash data analysis tools to:
    - Understand how the severity of injuries and survivability vary with age (to "normalize" injury criteria), and
    - Identify integrated safety mitigation options tied to specific applications of advanced plastics and composites.

2. Develop improved predictive crashworthiness tools and databases to enable utilization of plastics and composites in PCIVs such as:
    - Stronger passenger compartment designs with frontal crush boxes;
    - Roof designs and materials that are rollover proof;
    - Padded seating systems integrated with weight and impact force sensors;
    - Improved restraints to restrict side head movements and limit head and neck injuries; and

- Adaptive smart seat belts restraint systems "tuned" to occupant size, weight, and age or fragility.

3. Review NHTSA safety standards and cooperate with standards organizations in developing new standards for testing plastics and composites, to remove any bias and enable PCIV deployment.

4.3.2 Mid-term priorities (5-10 years) for Test and Evaluation

1. Verify and validate predictive engineering tools for:
   - Structural panels and interior components;
   - Prototype PCIV designs;
   - Alternative materials options; and
   - Performance in realistic crash scenarios.

2. Evaluate and quantify the PCIV safety benefits due to the application of:
   - Specific composite materials;
   - Advanced safety appliances based on plastics and composites; and
   - Improved models of PCIV crashes with larger, heavier vehicles.

4.3.3 Long-term priorities (10-15 years)

1. Partner in cooperative safety research related to PCIV technology integration.
2. Demonstrate and evaluate PCIV crash safety performance.
3. Participate in partnership research and demonstration efforts to enhance PCIVs crash safety for older occupants.

5. SAFETY ROADMAP FOR FUTURE PCIVs

5.1 The 2020 PCIV Safety Roadmap Development Strategy

The Senate Report 109-293 guidance to NHTSA was to "facilitate a foundation of cooperation…for the development of safety-centered approaches for future lightweight automotive design." Therefore, this report was designed to lay the foundation for future research cooperation, by conducting outreach to a broad cross-section of well-informed and diverse stakeholders on PCIV safety issues and goals. Their inputs on knowledge gaps, research priorities, and strategies to address the PCIV development and safety challenges and opportunities, as summarized in Chapter 4, provide the basis for consensus building on PCIV safety goals and milestones that can be translated into Safety Roadmaps. This process and resulting work product could also serve as a foundation for future public-private research partnerships and collaborative efforts with NHTSA centered on PCIV safety benefits, as desired by the Congress. At the same time, this research prioritization and safety-related roadmapping effort can inform and support other automotive technology roadmaps for emerging lightweight and fuel-efficient vehicles, to ensure that they too include a safety-centered component as part of their agenda.

The situation analysis, trends, and best practices discussed in Chapters 3 and 4 indicate that the automotive industry and their materials suppliers have both the expertise and great latitude in the adoption of composites in emerging and future fuel efficient and crash-safe vehicles, if affordability, reliability and manufacturability barriers can be overcome. This is shown also by the increasingly diverse uses of advanced composites in current potentially global compact concept cars (such as the 2007 GM Chevrolet VOLT), and in high-end sports cars, which feature advanced safety applications.

It is likely that a global platform PCIV will emerge by 2020, when competing material choices and structural designs will be optimized, in order to achieve desired economies of scale and secure market share. At present, competing light-weighting materials under investigation include aluminum, magnesium, and titanium alloys; hi-performance-steel; carbon fiber and other composites; and metal-plastic hybrids.

Public-private partnerships like USCAR are currently engaged in pre-competitive R&T efforts intended to overcome remaining barriers to commercial development and deployment of PCIVs, and are making progress towards that goal.

Crash safety is an implicit requirement for current FreedomCAR RDT&E efforts, no matter which materials and designs will be used (see for example the R&D program goals for automotive plastics and composites in Figure 5-4). Of the nine USCAR R&D consortia discussed in Section 3.3.2, at least four conduct research relevant to crash safety.

Although a comprehensive approach to crash safety has not been an explicit or central goal of existing technology roadmaps, it has been addressed through research projects conducted by the DOE/USCAR Crash Energy Management Working Group, and the Automotive Lightweight

Materials (ALM). If completed, the results could be used by the global automotive industry to cost-effectively achieve the 50-percent weight reduction and doubling of fuel efficiency goals by 2020.

Since the ACC Plastics Division's Technology Roadmap has not explicitly outlined the milestones and strategies focused on vehicle safety, this NHTSA-sponsored project complements it by explicitly addressing PCIV crash safety research needs, and provides a consistent timeline and milestones for achieving them. Furthermore, this project builds on existing roadmaps and leverages ongoing R&T partnership efforts, to cost-effectively capture and extend their potential safety benefits for future PCIVs.

As understanding of plastics and composites performance in structural safety applications improves, there will be no need to overdesign with plastics for crash safety assurance. Therefore, the materials and designs selected for vehicle shells will have to be both stronger and thinner, and therefore potentially cheaper and lighter.

5.2 Building on Existing Roadmaps for PCIV Safety Roadmap

5.2.1 The DOE/USCAR and FreedomCAR Roadmaps

The Federal government traditionally has played a key role in the pre-competitive Research, Development, and Demonstration (RD&D), with industry assuming the key role in subsequent commercialization of materials and technologies. The R&T roadmaps developed by the FreedomCAR teams are synergistic with the PCIV safety technology integration and deployment goals to some extent, and may serve as a reference point for some aspects of the 2020 PCIV safety roadmap.

The future 2020 PCIV concepts and vehicle technologies will support the ongoing development of multiple fuel-efficient design and innovative propulsion solutions, all of which would use strong and lightweight plastic and composite materials (hybrid electric, advanced ICE, flexi-fueled, fuel cell vehicles, etc.). It is assumed that all components, assemblies, and parts will have undergone safety verification and validation (V&V), as will the integrated PCIV vehicle system. Therefore, a "production-ready PCIV" based on "safety-centered design" principles is really a "system of systems."

In a critical review of the FreedomCAR and Fuel Partnership program, the National Academy of Sciences[60] 2005 report concluded that vehicle safety has not been addressed sufficiently, although it is a crosscutting issue of critical importance to successful transition to commercial hydrogen-fueled vehicles. It recommended (on pp. 8-9) that "DOE should form a new, crosscutting safety technical team"… and that "Both DOE and NHTSA need enough resources to carry out their assigned safety roles." Furthermore, the National Research Council emphasized NHTSA's lead role in a technical team dedicated to ensuring crash safety of future hydrogen

---

[60] National Research Council. (2005). *Review of the Research Program of the FreedomCAR and Fuel Partnership-First Report.*

vehicles, which should absorb the existing Safety Codes and Standards development team (pp. 36-37, ibid.)

R&D programs related to next-generation Hydrogen and Fuel Cell Vehicle Technology (FCVT) are already focused on advanced propulsion system development and safety. Advanced composites used in lighter propulsion systems and chassis PCIV applications will also improve safety by redistributing the weight for greater stability and for rollover prevention. Advanced plastics for lighter tops and glazing materials for integrated moon-roof assemblies would also contribute to improved mass distribution and rollover stability, as well as confer greater strength to the passenger "safety cage."

Although they do not explicitly focus on crashworthiness, most existing research and technology roadmaps overlap and address to some extent the knowledge gaps and R&D priorities identified by the experts surveyed (see Chapter 4). Expanding the knowledge base through research is necessary, but not sufficient to fully explore the safety of future PCIVs.

For over a decade, DOT participated in the PNGV and FreedomCAR partnerships that produced automotive technology roadmaps that included development and integration of advanced high-strength, lightweight materials (ALM) into the next generation of vehicles, such as the referenced:

- Technology Roadmap for 21st Century Truck Program (2000);
- National H2 Energy Roadmap (2002);
- FreedomCAR and Fuel Partnership: Materials Technology Roadmap (2006);
- Hydrogen Posture Plan- An Integrated RD&D Plan (2006); and
- FreedomCAR and FreedomFUEL Partnership: Materials Technology Roadmap (2006).

Safety appears to be an implicit goal for all USCAR R&T team efforts, which can be leveraged as "building blocks" for future PCIV safety-oriented designs and development, as illustrated in excerpted Figures 5-1 through 5-5:

- Figure 5-1 shows the general Automotive Composites Consortium (ACC) timeline for developing new automotive materials and technologies, which set a commercialization decision milestone in 2015. It is consistent with the PCIV safety roadmap and NHTSA Technologies integration timetable shown in Figure 5-7.

- Figure 5-2 excerpted from the Materials Technologies Roadmap, shows activities focusing on diverse Automotive Light-weighting Materials, as well as on high-strength materials for structural applications to the entire fleet of light and heavy duty vehicles, and on specialty high-temperature materials for propulsion systems.

- Figure 5-3, reproduced from the recent FreedomCAR Vehicle Systems Analysis Technical Team (VSATT)[61] Roadmap, shows on the left side that the strategy for integrating emerging lightweight, high-strength materials and technologies into a future

[61] FreedomCAR and Fuel Partnership-Vehicle Systems Analysis Technical Team, (September 2006).

reference fuel-efficient vehicle is congruent with the R&D priorities identified in Chapter 4. These goals include: modeling and simulation for a "reference vehicle" (including a PCIV), benchmarking the performance of components and the full vehicle, as well as Verification and Validation (V&V) prior to technology integration into commercial production vehicles. The USCAR Vehicle Systems Analysis Technical Team (VSATT) has not yet defined the 2010 reference vehicle, but it plans to integrate the results from all other USCAR teams, use advanced modeling and simulation tools, and conduct validation testing, benchmarking and field demonstrations of components and system fleet performance of concept vehicles.

- The FreedomCAR Materials Technology Roadmap specifies strategic research goals and objectives for lightweight automotive composite materials essential to enabling PCIV safety, as shown in Figure 5-4. In particular the FreedomCar Roadmap focuses on alternative light-weighting materials, including Carbon Fiber/Polymer Matrix Composites, Glazing materials, as shown in Figure 5-4a and 5-4b, respectively.
  - The USCAR Carbon Fiber/Polymer Matrix Composites (CF/PMC) materials roadmap shown in Figure 5-4a includes as an explicit Long-term goal (5-10 years): "Predictive Modeling for Crashworthiness" and "Optimum design and prediction of durability and failure."
  - Similar structural performance and durability goals are explicit research objectives for the Glazing plastics roadmap (Fig 5-4b).

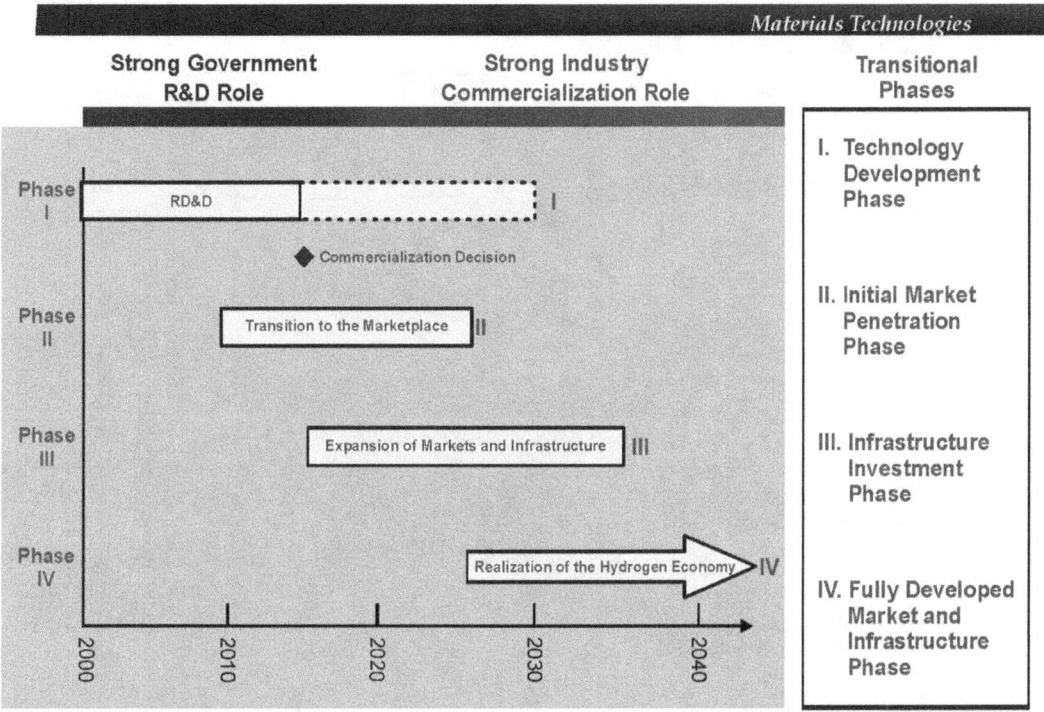

Figure 5-1: Timeline, performers, and roles for the DOE FreedomCAR Automotive Composites Consortium partnership for materials technology development. Source: The DOE/USCAR Technology Roadmap for lightweight, energy-efficient vehicles (from the presentation by J. Carpenter, May 2006 SAE/Industry)

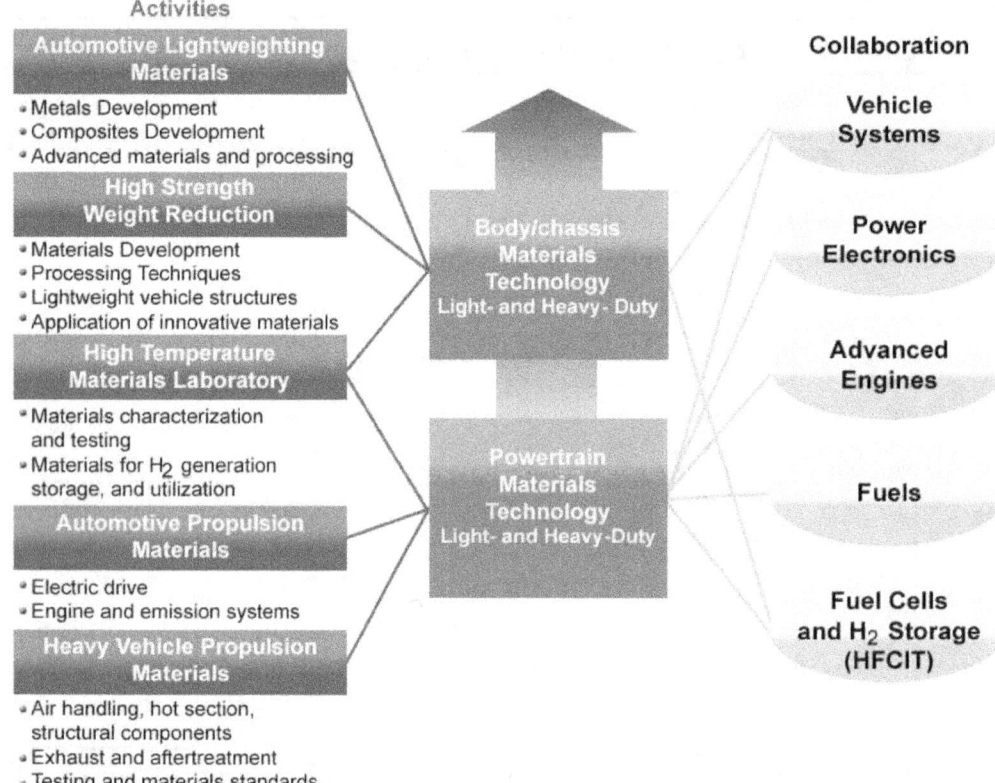

Figure 5-2 : Lightweight and high-strength materials for applications to structures and propulsion in next-generation vehicles are an explicit objective under the FreedomCAR and Fuel Partnerships and the 21st Century Truck Partnership.

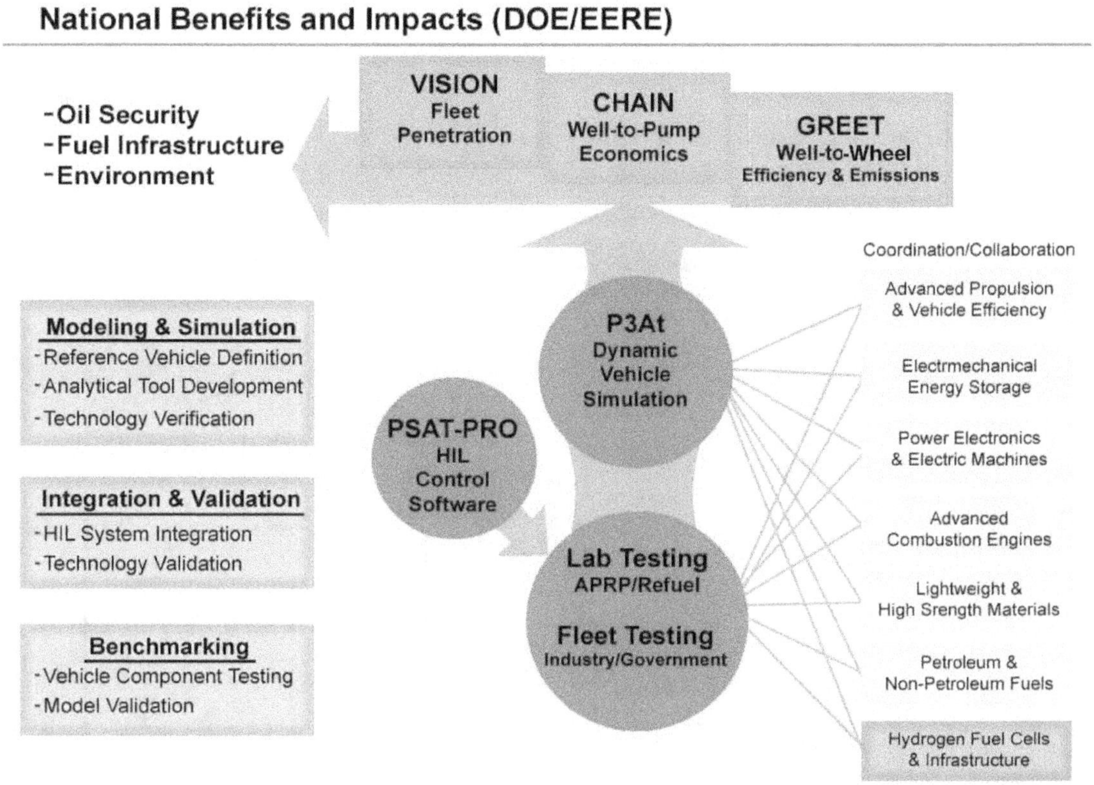

Figure 5-3: Depiction of the USCAR strategy to model, develop, integrate, verify, and validate emerging lightweight, high-strength materials and technologies for a future reference fuel-efficient vehicle. (Reference: "FreedomCAR and Fuel Partnership-Vehicle Systems Analysis Technical Team," September 2006)

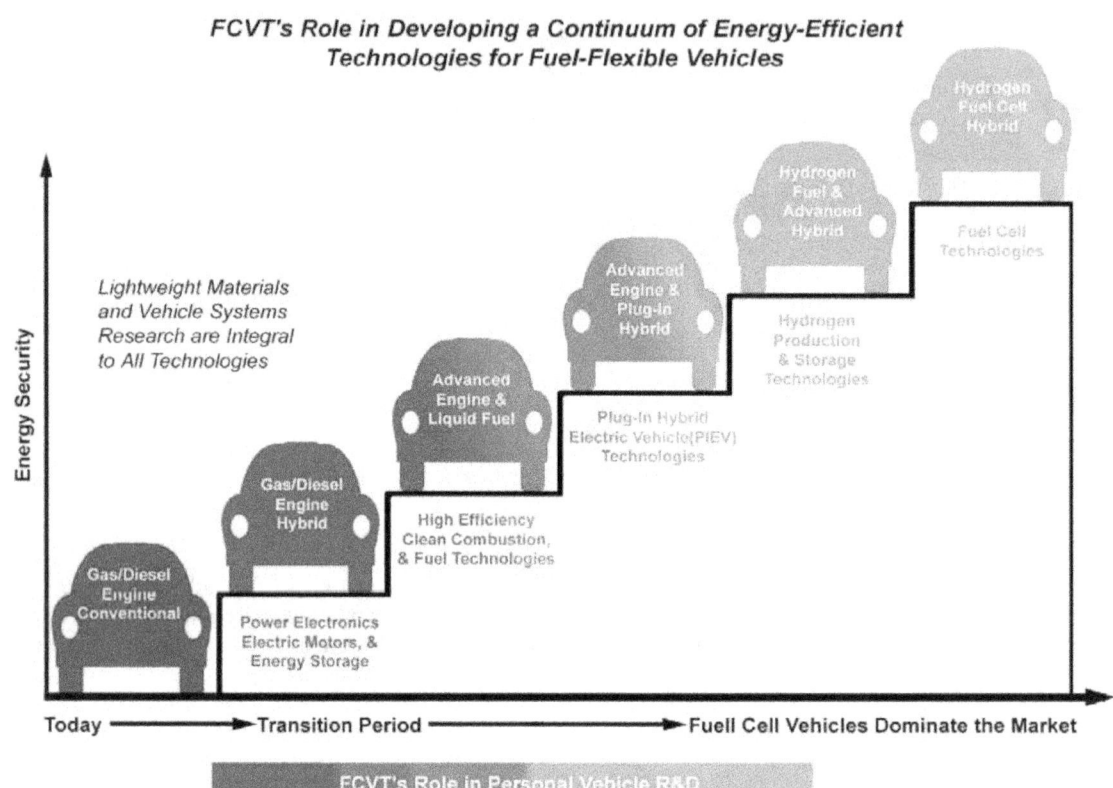

Figure 5-4: Figure from "Driving Technology: A Transition Strategy to Enhance Energy Security," DOE/EERE May 26, 2006. Safety is an implicit "utility" in every step on the ladder depicting the progress towards energy-efficient FCVs over time.

Carbon Fiber/Polymer Matrix Composites — Materials

Strategic Framework	Research Objectives	Current Projects and Perceived Needs
Carbon FRP developed to replace steel parts will achieve a mass reduction og 50-60%. Today's relative cost per part vs. steel is 2.0-10.	**Long-term goal (5-10+ years)** Enable the use of carbon FRP to create structures, relative to incumbent material, with 55-65% wt reduction and cast parity. - Low-cost carbon fiber - Predictive modeling for crashworthiness - Robust joining technology including optimum design and prediction durability and failure - Economical recycling **Intermediate-term goals (3-5 + yrs)** Demonstrate technical feasibility of the manufacture of large structures from automotive grade carbon FRP so as to replace incumbent materials. -Develop durability-drive design guidelines -Evaluate viable new molding processes -Guide suppliers in developing carbon fiber materialsto suit automotive requirements	**Current Projects** • ACC040: Development and optimization of chopped carbon fiberpreforms • ACC115: Optimize injection-compression processing based on characteristics properties of the preforms • Predictive modeling of structural adhesive joints • ACC100L Developing models for prediction of crash energy management. • ACC080: Focal Project 3 Design and manufacturing validation of a carbon fiber intensive BIW aimed at > 60% mass reduction. • ACC205: Low Cost Test Methods for Advanced Automotive Composite Materials: Creep Compresion Fixture **Topics RequiringFurther Research** • Structural thermoplastic composites • Class-A structural compsites

Technology-Specific 2010 Goal for Materials: To enable lightweight vehicle structures and systems, the goal is: Material and manufacturing technologies for high volume production vehicles which enable/support the simultaneous attainment of: *1.) 50% reduction in the weight of vehicle structure and subsystems, 2.) affordability, and 3.) increased use of recyclable/renewable materials.*

Figure 5-4a: Selected Roadmaps for automotive composite materials from "FreedomCAR and FreedomFUEL Partnership: Materials Technology Roadmap" (October 2006). See Carbon-Fiber/Polymer matrix composites at www1.eere.energy.gov/vehiclesandfuels/pdfs/program/materials_team_technical_roadmap.pdf

Glazings

Strategic Framework	Research Objectives	Current Projects and Perceived Needs
Glazing represents approx. 5% of the mass of a typical automobile. Lightweight glazing alternatives can result in a 50% weight reduction compared to conventional materials. Today's relative cost per part vs. conventional (laminated glass) material is 5x. Glazing can also contribute to reducing vehicle energy consumption while improving passenger comfort by reducing cabin thermal load.	Long-term goal (5-10+ years) Use lightweight glazing materials for 50% Wt reduction with cost parity to today's materials and no compromise in safety performance (as defined under Intermediate Goals) Intermediate goals (3-5 yrs.) Use lightweight glazing materials for 30% wt. reduction with cost 2x vs. today's materials. A. Develop low cost lightweight glazing that will provide optical clarity and image quality, penetration resistance, and reduce potential for laceration and head injury B. Develop materials with comparable structural and durability performance C. Develop modeling and simulation methods to predict noise and thermal transmission and structural performance D. Develop materials or techniques to reduce noise transmission by 6 db E. Develop materials to reduce cabin heat load by 95% IR light blocking	Current Projects PNNL - Lightweight and High Performance Alternatives for Future Automotive Glazing Systems - Evaluate thinner glass, glass/plastic laminates and injection molded interlayers - Investigate design alternatives such as glazing curvature variations and laminates to reduce noise transmission Topics Requiring Further Research •Develop materials or techniques to reduce cabin heat load • Develop materials, designs and processes for targeted db noise reduction •Develop models to predict heat load, sound transmission, impact resistance. • Use models to identify and drive continuous improvement efforts

Technology-Specific 2010 Goal for Materials: To enable lightweight vehicle structures and systems, the goal is: Material and manufacturing technologies for high volume production vehicles which enable/support the simultaneous attainment of: 1.) 50% reduction in the weight of vehicle structure and subsystems, 2.) affordability, and 3.) increased use of recyclable/renewable materials.

Figure 5-4b: The structural strength of glazing plastics and composite materials for vehicle windows and roofs are very important to occupant safety in crashes (e.g., to prevent ejection in rollovers).

5.2.2 The NHTSA Integrated Safety Strategy and Timeline

In light of NHTSA's interest in crashworthiness (passive safety) and crash avoidance (active safety), this Roadmap focuses on the structural safety of advanced composite materials, including thermoset and thermoplastic polymer matrix composites and newer nano-composites, that could be used in future PCIVs. The Roadmap highlights the key enabling research areas needed to fill existing knowledge gaps and their probable timetable.

Recently, NHTSA adopted an "integrated safety" approach to vehicle crash safety research, in order to reap the safety benefits from the synergistic deployment of multiple safety technologies. These are depicted in Figures 2-1 and 2-2 and discussed above in Section 2.1.4, PCIV 2020 Vision. This approach is intended to better prevent crashes, manage the impact shock, mitigate injury severity, and enhance occupant protection through the integration of active and passive safety technologies. In addition, the NHTSA safety research program has developed a safety technology integration strategy and timeline, shown in Figure 5-7. This safety research strategy will capture the benefits of industry innovations and inform future regulatory development. This research and technology strategy cuts across all phases of Haddon's Matrix for crash injury prevention, event response and injury mitigation, and recognizes the complex interactions of vehicle performance, driver behavior and the environment.

As shown in Figure 5-7B, NHTSA's strategy is to monitor and quantify the safety benefits of deploying and integrating the broad range of innovative active and passive occupant protection systems. The technologies listed include improved energy absorbing structures, and comprehensive and adaptive restraint systems, in which advanced composites and other high-strength materials and structures envisioned for PCIVs could play a key role.

In the mid-term, by 2014, NHTSA envisions capturing the safety benefits of deploying active safety technologies for Driver Warning and Assistance Systems, ranging from the currently mandated Electronic Stability Control (ESC) to Automatic Braking and Collision Notification systems. Several driver assistance technologies appear particularly promising for enhancing older drivers' and passengers' protection, such as night-vision systems, lane departure warning, and intersection collision avoidance systems.

This crosscutting, integrated safety research strategy could be extended to embrace the application of advanced materials in structural safety applications, and any complementary safety enhancements to future PCIVs.

5.2.3 Extending the ACC-PD Automotive Technology Roadmaps to PCIV Safety

The American Chemistry Council – Plastics Division's automotive technology roadmap for plastics[62] treated safety (see Chapter 3) as an integral part of the "Safety, Environment, and Health" issue. It noted, however, the rapid progress in passive and active safety features dependent on plastics improvements, and improvements in crash energy management (CEM)

---

[62]See *Plastics in Automotive Markets Vision and Technology Roadmap*, American Plastics Council, 2001.

through use of advanced structural materials, crumple zones, and design strategies. The ACC-PD roadmap's coverage of Advanced Materials Systems (in Ch. 4, ibid.), as well as the Predictive Engineering, Testing, and Modeling coverage (Chapter 5, ibid. reproduced in Figure 5-7) and strategic objectives are very similar to the USCAR Roadmaps discussed above. Furthermore, there is broad consensus between the ACC-PD roadmap and the RDT&E priority needs that resulted from the experts' survey in Chapter 4. This broad-based consensus suggests that vehicle safety will be able to benefit from, and build on the progress achieved by ongoing partnership programs in the next decade.

The American Chemistry Council Plastics Division (ACC-PD) provided the following guidance regarding the definition of a "baseline" 2020 PCIV and probable milestones and progress metrics relevant to a PCIV Safety Roadmap:[63]

- "Through PCIV designs, cars will be on the average 50-percent lighter than today's passenger automobiles."
- "A PCIV will include multiple materials in its construction…Major automotive systems are at least partially designed and manufactured in plastics and composites."
- "PCIVs incorporate all light vehicles including passenger cars, SUVs, and pick-ups, and have the potential to embrace medium and heavy-duty vehicles as well."

The ACC-PD also offered the following clarifications regarding the probable PCIV technology readiness level (TRL) by 2020, to reflect on whether the PCIV would be only in the conceptual design stage by 2020, or technology prototypes in the Test and Evaluation (T&E) phase, or if PCIVs will actually be commercially deployed:

- "At least one manufacturer will introduce a production PCIV by 2020 acceptable to consumers, which meets applicable Federal Motor Vehicle Safety Standards (FMVSS)."
- "In addition, all major vehicle manufacturers will be PCIV design-capable by 2020."

Continuing improvements in traditional vehicle interior plastics applications will enable future PCIVs to feature:

- An improved "safety cocoon" in the passenger cabin to better protect older occupants;
- More effective interior padding, such as a roof canopy to protect occupants in rollover crashes;
- Improved, adaptive seat assemblies that cradle the passengers in impacts; and
- Advanced, smart safety appliances able to cushion and attenuate impact forces, and to redistribute or deflect impact loads.

Since plastics and composites are already used to a large extent for interior safety applications (air bags, foam padding, seat assemblies) and industry continues to improve them incrementally, the major growth opportunities for future PCIVs are in exterior and propulsion system applications.

---

[63] American Chemistry Council Plastics Division (ACC-PD), *PCIV Automotive Safety Roadmap Input for the Volpe Center*, March 19, 2007. E-mailed by Cole & Associates, Inc.

5.3 Translating R&D Priorities into PCIV Safety Roadmaps

The RD&T priorities identified for each time horizon in Chapter 4 could be met progressively over time and integrated seamlessly into best industrial practices in order to develop a production-ready commercial PCIV prototype fully compliant with NHTSA safety requirements and industry Best Practices:

- In the near-term (3-5 years): address materials engineering knowledge gaps (standardized test procedures, databases, modeling and simulation tools) and identify effective protection strategies for older occupants.
- In the mid-term (5-10 years): develop PCIV designs with >50 percent plastics and plastic composites content, then Test and Evaluate (T&E) components and subsystems to model integrated system crash performance and benefits.
- In the long-term (15-20 years): document, analyze, verify, and validate the effectiveness of the integrated crash safety strategy for PCIVs. Another specific long-term goal is to demonstrate superior PCIV crash safety protection of older drivers and occupants, to better advance PCIV public acceptance and market penetration.

The present report and PCIV Safety Roadmap are intentionally general and designed to offer a high level status report on current R&D, knowledge gaps on materials safety performance and to guide future national RDT&E activities and milestones. The new safety Roadmap schematics in Figures 5-8, 5-9 and 5-10 preserve the ACC-PD's Technology Integration Report format, as well as extend and complement it. If addressed by future R&D efforts, the Roadmap could serve to fill existing knowledge gaps concerning the predicted and actual crash safety performance of composite materials in specific PCIV designs (as shown in Figure 5-9, Strategic PCIV safety priorities and in Figure 5-10). Therefore, this initial effort gives industry freedom in innovative design, choice of materials and of safety technology solutions. Follow-on collaborative and coordinated research is needed to further develop the safety roadmap in greater depth and detail, to evaluate potential safety benefits from specific materials and technology applications.

5.4 Potential NHTSA Role in Safety Assurance of Future PCIVs

In the near- and mid-term, industry, academia, and professional standards organizations will continue to play a major role in developing and implementing the PCIV safety roadmaps, with the cooperation and funding support of multiple Federal agencies.
Road safety and consumer stakeholder organizations, as well as NHTSA, as a safety regulatory agency, will have important roles in the long-term facilitation and acceptance of PCIV as safe consumer market products. Prototype PCIV crash safety will have to be modeled, tested, and evaluated under realistic field-conditions, to verify their compliance with NCAP and FMVSS regulations, and perhaps improved for large scale commercial deployment and consumer acceptance beyond 2020.

NHTSA's role in influencing a safety-centered future PCIV will be consistent with its mission as a vehicle safety regulatory agency. NHTSA's ability to participate in public-private R&T partnerships and to cooperate in development of technical standards applicable to PCIVs would depend on agency policy and priorities, as well as on resources. In the near and mid-term, NHTSA could continue to monitor and evaluate the findings of FreedomCAR technical groups focused on vehicle safety and occupant protection.

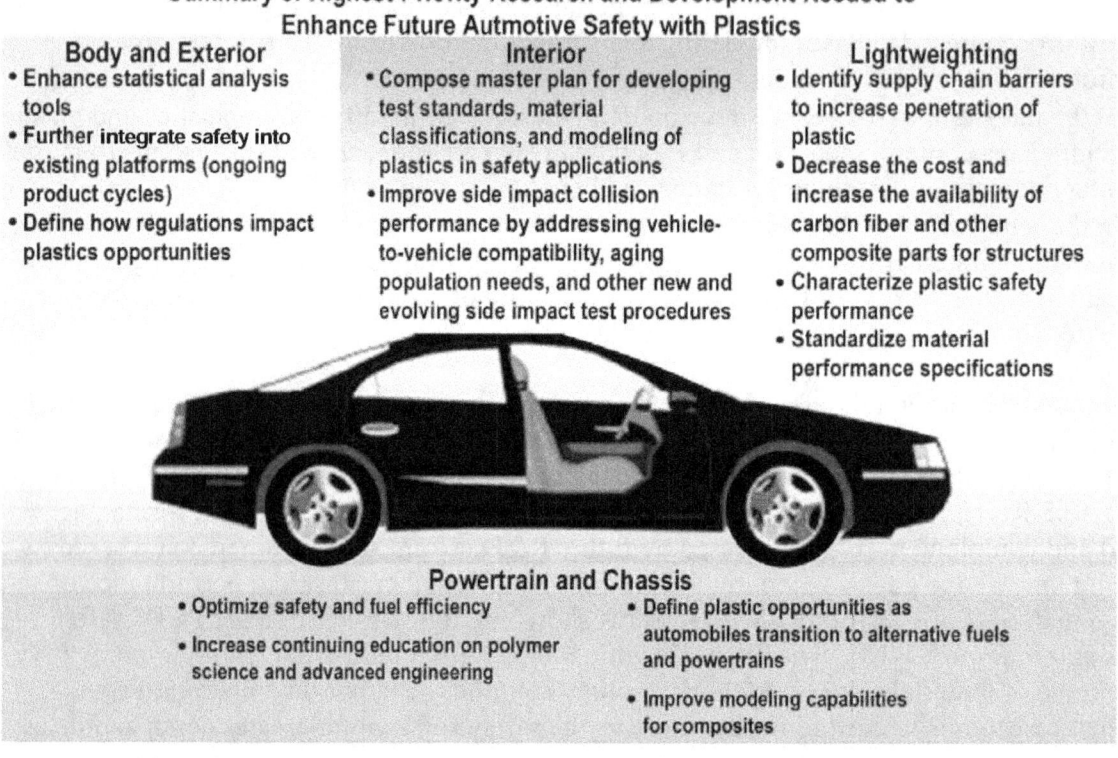

Figure 5-5: Summary of research and development priorities for plastics and composites in automotive safety applications, as identified in the May 2006 ACC-PD report.

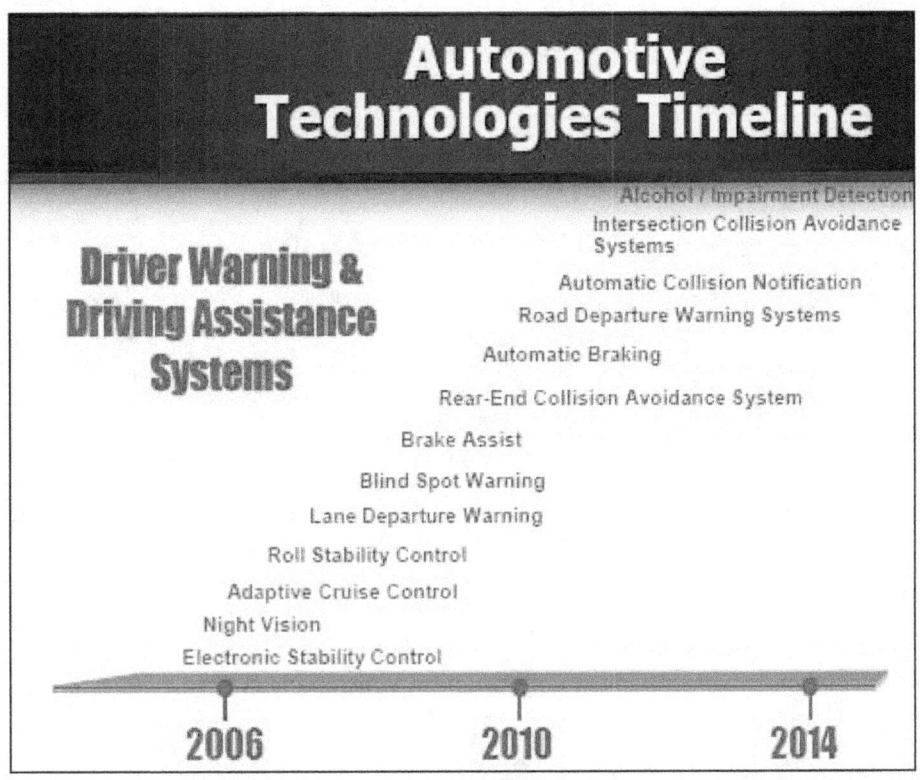

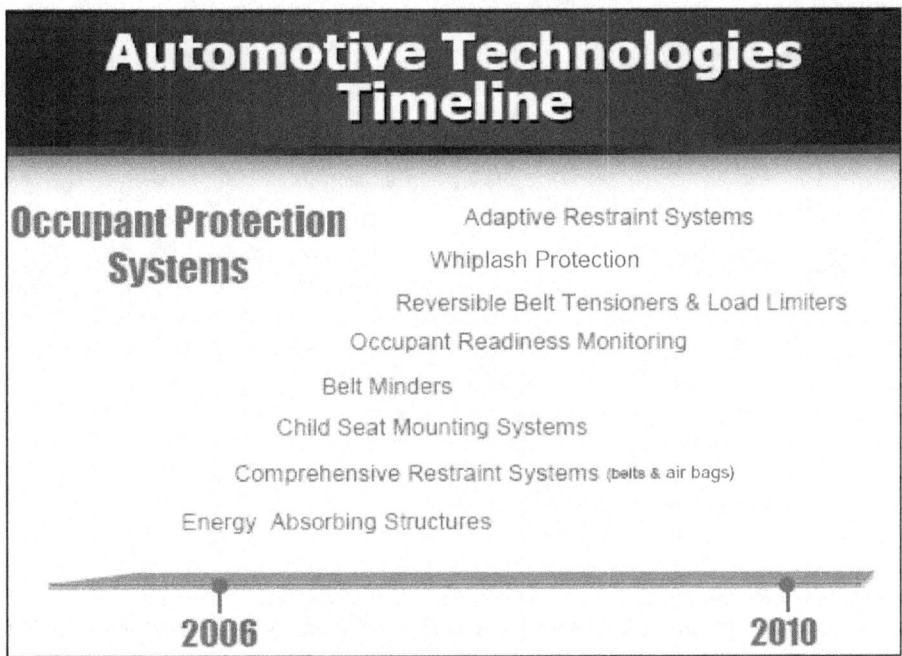

Figure 5-6: A and B- NHTSA technology integration timeline for active and passive safety technologies, from "Overview of NHTSA Research for Enhancing Safety," October 11, 2006. Detroit presentation by Dr. W. T. Hollowell.

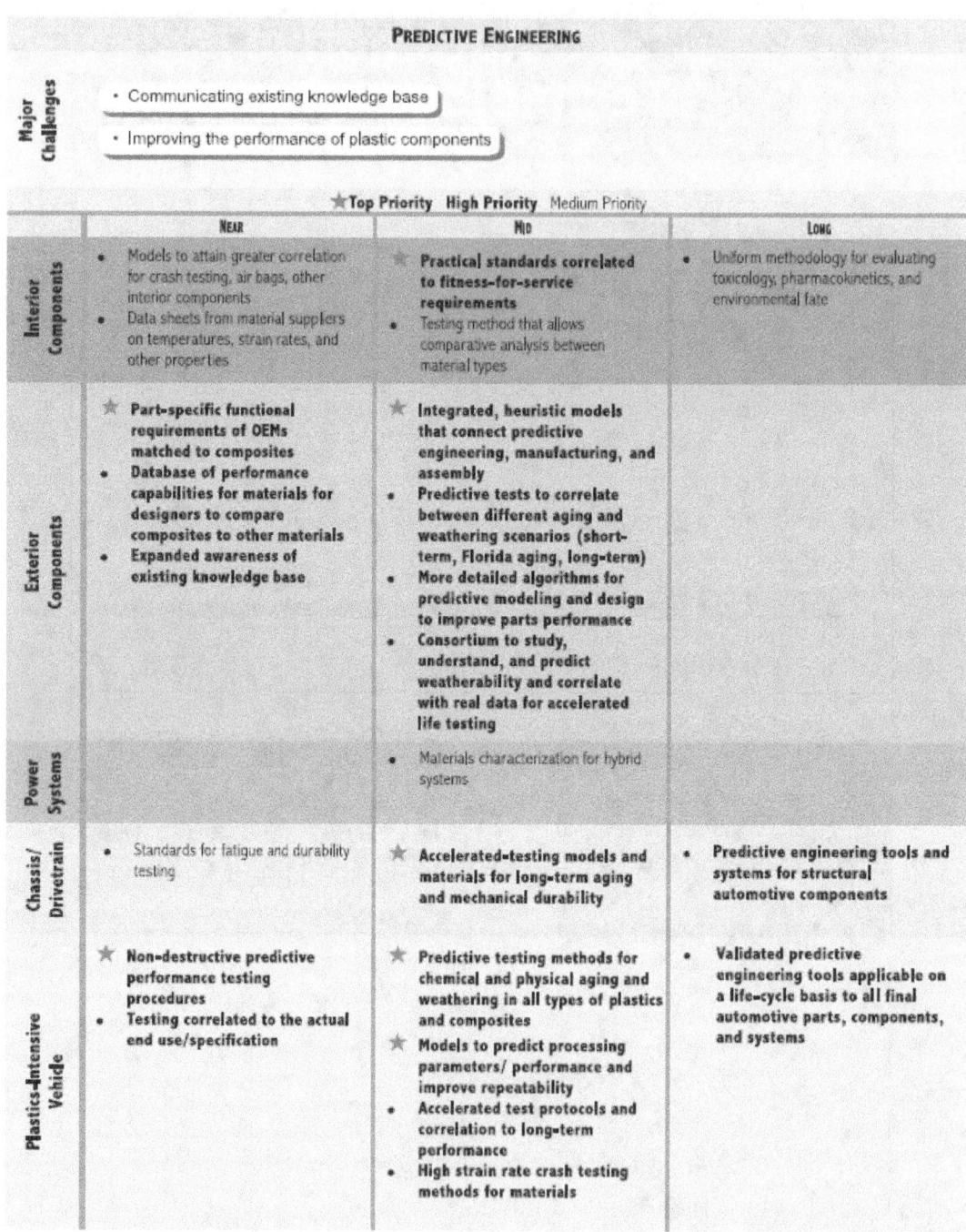

Figure 5-7 : Roadmap for Predictive Engineering tools needed to model crash safety performance of plastics-intensive vehicles reproduced from the 2001 ACC-PD report "Plastics in Automotive Markets-Vision and Technology Integration." This is a key mid-term (5-10 years) research and technology priority identified in this study.

RESEARCH AND TECHNOLOGY INTEGRATION FOR PLASTICS AND COMPOSITES INTENSIVE VEHICLES (PCIV) SAFETY

SCREEN & SELECT COMPOSITE MATERIALS FOR PLASTICS AND COMPOSITES INTENSIVE VEHICLES LIGHT-WEIGHTING DATABASE TAILORED TO PART DESIGN AND FEATURES.

Present	Near-term (3-5 years)	Mid-term (5-10 years)	Long-term (10-15 years)	Far-term (15+ years)
	Use Computational Crashworthiness Model to Predict Loading, Crash Energy Management, and Failure Behavior.	Prototype Component or Sub-System (e.g. Door Panels)	Integrate Into Full Scale Concept Car, Field Test, and Compare Safety Performance in Standardization Tests	NHTSA, DOE and P3 Review NCAP PCIV Crash Safety Results
	Compile a Comprehensive Database on Mechanical Properties (Static & Dynamic)	Perform Accelerated Validation & Verification Static And Dynamic Lab Scale Test to Optimize Crash Performance	Develop Test Procedures	
	Optimize Design for Crash Safety Performance, focusing on Older Demographic		Field Test PCIV Prototypes to Improve Production Model	Collect and Analyze Crash-Test Data to Verify Older Occupants Safety

Figure 5-8: Timeline for Research and Technology Integration for PCIV Safety Roadmap

SAFETY ROADMAP FOR FUTURE PCIVS

Enhancing Plastics and Composites Intensive Vehicles Safety Performance With Plastics

Challenges / Milestones
- Design concepts
- Materials screening
- Testing standards
- Simulations and validation
- Systems integration
- Crash safety testing
- Performance metrics
- PCIV deployment

Performers	Near-term (3-5 years)	Mid-term (5-10 years)	Long-term (10-15 years)
Industry – Government – University Public Private Partnerships (P3)	Research, Development, and Technology on Automotive Composites	Test & Evaluation Of PCIV Prototype Crash Safety	System Integration of PCIV Safety Technologies 10-15 yrs.
	Develop Testing Standards and Safety Evaluation Tools for PCIV Designs 1 yr.	Crash Safety Verification & Validation Simulations 5-6 yrs.	
	Select Lightweight Structural Materials for PCIV 2-3 yrs.	Crash Safety Testing & Validation For PCIV Subsystems and Vehicle 6-8 yrs.	PCIV Commercial Deployment 15+ yrs.
	Develop PCIV Materials Processing / Parts Fabrication 3-5 yrs.		
NHTSA Role	NHTSA Monitors Progress in Crash Safety Research and Development	NHTSA Evaluates Results of Crashworthiness Verification & Validation	NHTSA Verifies PCIV Crash Safety Compliance (NCAP and FMVSS)

Figure 5-9: Roadmap to PCIV Safety

SAFETY ROADMAP FOR FUTURE PCIVS

Strategic Priorities for 2020 Plastics and Composite Intensive Vehicles Safety Assurance

Major Challenges: Perform focused, coordinated and integrated safety research, development, test, and evaluation Program to improve crash-safety performance of plastic/composite components and subsystems for a 2020 PCIV.

Research, Development, Test, and Evaluation Priority

	Near-term (2007-2010)	Mid-term (2010-2015)	Long-term (2015-2020)
Materials Selection	Perform research and technology to address "knowledge gaps" on crash-safety performance of PC materials	Modeling and simulation to verify and validate plastic/composite crash safety in structural, semi-structural applications	Demonstrate integrated safety performance for prototype PCIV to enable commercial deployment
	• Standardize testing protocols for composite materials	• Validate plastic/composite materials choices in safety applications	• Industry crash-test and self-certify PCIV safety
Testing Crash Performance	• Characterize mechanical behavior of plastic/composite materials in safety applications	• Prototype and test components (door panels, roof, front and back "crush boxes")	• Identify and overcome PCIV crash-compatibility problems for all occupants
	• Establish comprehensive Database for light-weighting materials options	• Verify and validate for baseline PCIV design to evaluate integrative safety system performance	• Demonstrate enhanced PCIV safety performance for older occupants (using advanced dummies)
PCIV Integration	• Refine predictive engineering tools for Modeling and Simulation of PCIV components and system crash performance	• Devise and evaluate special crash-protection needs for older occupants	• NHTSA verifies PCIVs compliance with crash- safety regulatory requirements
Milestones		↑ • Test Standards Issued • CHM-17 Crash Energy Database Standardized and Current • Modeling and Simulation Crash Safety Tools Available	↑ • PCIV Structural and Propulsion Validated for Realistic Crash Loads • Full PCIV System is Crashworthy • Improved Older Occupants Survivability is Demonstrated

Figure 5-10: Strategic Priority Activities Leading to PCIV Safety Assurance by 2020

6. Appendices

6.1 Appendix 3.1- PRIORITY CROSSCUT SAFETY ISSUES IN THE APC WORKSHOP REPORT *ENHANCING AUTOMOTIVE SAFETY WITH PLASTICS*

1. Accommodate Changing Demographics - Older Population Growth

- Increasing vehicle side impact compatibility
  - Development of "Smart Plastics"
  - Optimize energy management for side impact
  - Design side air bags to meet new standards
  - Improve rollover safety

- How to accommodate older population and other diversity issues
  - Optimize energy absorption and transfer
  - Develop energy-absorbing plastic structure covering door-to-door
  - Develop a practical elderly test dummy
  - Develop age-based air bags and performance metrics
  - Develop appropriate lower-body countermeasures
  - Improve effective part design for energy management
    - New materials
    - Inflatable air bags
  - Define Material energy absorption characteristics
  - Develop inflatable seat belts for the elderly
  - Consider multi-stage air bags
  - Develop belt materials that can uniformly transfer occupant loads

- Providing safety solutions in reduced package spaces
  - Develop inflatable padding for rollovers
  - Develop unconventional air bags
    - Smart air bags

2. General safety priorities
- Further integrate safety into existing platforms (ongoing product cycles) [body and exterior]
- *Compose master plan for developing test standards, material classifications, and modeling of plastics in safety applications [interior]
- Characterize plastic safety performance [light-weighting]
- Standardize material performance specifications [light-weighting]
- Optimize safety and fuel efficiency [Powertrain & Chassis]
- Improve predictive modeling capabilities for composites. [Powertrain & Chassis]
- Enhance crash performance with improved energy management

- Accommodate changing demographics to older population
- Plastics can contribute to vehicle compatibility through light-weighting, structural interactions, and energy sharing.
- Plastics can enable tighter integration of active and passive safety systems
- Plastics experts should observe full-vehicle crash testing to gain insight into real-world challenges.

2.1 INTERIOR
- Improving passenger impact protection
- Reducing passenger ejection
- Retarding fire and warning of potential problems
- Enabling safer driving
- Challenge: Improving passive safety
- Providing safety solutions in reduced package space
  - Smaller/lighter systems could allow for overall reductions in weight
- Increasing vehicle compatibility (Side Impact)[64]
  - Develop pre-crash sensing and adaptive restraints
- Providing safety solutions in reduced package spaces
  - Design to increase crush space
  - Develop integral seat design with functional fabric and pre-crash mechanism
- Competition between lightweight versus safety
  - Design and safety test integrated components
  - Define active safety capabilities and light-weighting impacts
- Improving seat belt design
  - Demonstrate new plastics potential
- Crosscutting issues:
  - For structural component, improve testing and understanding of fatigue, high-rate response, material aging effects
  - Update OEM specifications and test methods to target plastic properties or performance needed
  - Improve systems for grouping plastics
  - Understand aged material performance and design

2.2 BODY AND EXTERIOR
- Absorbing crash energy and improving stability
- Reducing head and body injuries
- Challenges: passive safety
  - Managing crash energy to protect occupants
  - Achieving weight reduction for body/exteriors without compromising safety
  - Improving roof crush performance in rollovers
- Challenges: validation
  - Lack of demonstrated durability and reliability data from materials providers

---

[64] Note that current vehicle incompatibilities challenge the ability to balance side impact, intrusion resistance, and energy absorption, but could be overcome in PCIVs by using a modular approach.

- Priority activities
  - Resisting vehicle intrusion and roof crush (demonstrate and validate)
  - Reducing body and exterior weight without negatively impacting safety
- Activities needed to enhance future automotive safety in the body and exterior
  - Vehicle intrusion and roof crush
    - Develop, test, and certify body panels as part of safety structure
    - Design foam interior structure.
    - Improve energy absorption in frontal or side impacts
    - Evaluate plastic-metal hybrid for roof crush
  - Develop prototype structure and performance data

2.3 POWERTRAIN AND CHASSIS
- Challenges:
  - Fuel efficiency
    - Light-weighting vehicles without compromising crash performance
  - Computational Methods
    - Limited predictive modeling capabilities for all types of polymer composites (testing is essential)
- Activities needed:
  - Optimizing Safety and Fuel Efficiency
    - Improve the understanding of safety factors influenced by light-weighting
    - Study trade-offs of parts integration
  - Expanding predictive modeling capabilities for all types of Polymer Composites
    - Develop FEA models and standard test methods
  - Responding to impacts of alternative fuel vehicles and power-train options
    - Develop non-flammable laminates and sandwich structures

6.2 Appendix 4.1- EXPERTS INTERVIEW GUIDE

1. Field of Expertise: Please indicate your professional background and affiliation (check box)
   - Federal ☐
   - Industry ☐
   - Academia ☐
   - Professional or Standards Association ☐
   - Nonprofit NGO ☐
   - Other ☐

2. Please indicate if and how you have been/are involved with automotive
   - safety applications
   - plastics and composites design and development
   - technology integration and deployment
   - vehicle crashworthiness and energy management (testing)

3. Which safety applications used plastics and composites (please specify material and check one or more, if applicable):

   - Structural strength (body chassis, motor) ☐
   - Interior foam ☐
   - Seat belt or ☐
   - Air bag system ☐
   - Other (windshield, glazing, roof, bumpers) ☐

   Notes (please discuss):

4. Please estimate how long it took in the past and would take in the future for automotive plastics and composites R&DT&E through technology integration for commercial deployment:

   - Screen for materials and select supplier
   - Develop and evaluate design using Modeling and Simulation
   - Perform Verification and Validation
   - Prototype and field- test (components or full scale vehicle?)
   - Develop material quality and manufacturing processes
   - Integrate the technology into commercial vehicles
   - Confirm and prove its value (benefits to cost ratio) over time
   - Pass the NHTSA crash tests for safety certification (FMVSS, NCAP, Voluntary standards and guidelines)

5. Please designate your Top 3 priority safety applications of plastics and composite materials for the:
   - near-term (3-5 years)
   - mid-term (5-10 years)
   - long-term (10-15 years)

   (Please match if possible a specific material to a specific safety application in a real vehicle model, and justify your recommendation).

6. Please comment on the crashworthiness performance of composites in existing vehicles and proposed PCIV and identify:

    i. Knowledge gaps

    ii. Research needs

    iii. Time to develop and demonstrate safety

    iv. Time to penetrate the market (in United States? Abroad?)

7. If a major objective is to enhance the protection of aging drivers and passengers in crashes, what are the most promising safety enhancements for PCIVs in your opinion?

8. How can the safety advantages of new materials/applications for everyone in general, and for aging drivers in particular, be proven both qualitatively and quantitatively?

9. What are the major knowledge gaps and technical challenges to greater use of advanced composites)?

10. Please identify any safety-related regulatory barriers and standards development needs to foster PCIV deployment.

11. How can DOT/NHTSA best facilitate and speed up the deployment of lightweight PCIVs, while ensuring they perform at equal or better safety?

12. Please suggest when and how should NHTSA be involved in the safety performance of automotive plastics and composites?

6.3 Appendix 4.2-LIST OF EXPERTS INTERVIEWED[65]

Name	Affiliation	Responses
Dr. Joseph Carpenter	DOE HQ	Phone interview, papers received
Rogelio Sullivan	DOE HQ	Discussion at ACCE06, phone conversation, presentation received
Dr. David Warren	DOE/ORNL	Phone interview, form completed
Dr. Mark Smith	DOE/PNL	Phone interview (DOE/APC/NSF CRADA)
Drs. David Weber and Ron Kulak	DOE/ANL	Phone contact, form completed and e-mailed
Dr. Steve Riddella	NHTSA HQ (formerly TRW)	Phone interview
Don Willke	NHTSA/VRTC	Phone interview
Felix Wu	DOC/NIST/ATP	Phone interview, web posted refs.
Prof. Joseph Coughlin	MIT Age Lab and CTS	Meeting, papers provided
Prof. Paul Lagace	MIT Aero/Astro Adv. Composite Lab	Phone interview, form and supporting information was e-mailed
Prof. Marc Ross	U. Mich./UMTRI	Phone interview, form completed and papers e-mailed
Dr. Susan Hill	UDRI	Phone interview, papers e-mailed
Prof. Paolo Feraboli	UW/Dept. of Aero. & Astro.	Phone interview, form completed and papers e-mailed
Dr. Jackie Rehkopf	Exponent, Chair SAE Plastics Stds. Devpt. Comm.	Phone interview, meeting at ACCE06, form completed
Richard Jeryan, Mark Botkin, Chaitra Nailadi	Ford R&D Ctr. GM R&D Ctr. DaimlerChrysler	Phone interview with ACC Crash Energy Management (CEM) Working Group (WG), e-mailed ACC Plastics issues
Dr. David Zuby	VP Res., IIHS	Phone Interview, papers provided
D. McLeod & team	Dow Automotive	Form completed, e-mail suggestions

---

[65] About a dozen other experts were approached by phone or e-mail and did not respond, including: NSF, AARP, and several industry experts identified by APC or literature search. The total number of non-Federal respondents is less than the 10 allowed by OMB PRA limitations on surveys. The CEM-WG group interview included several Big-3 and DOE labs representatives.

7. REFERENCES

1. Commission for Global Road Safety. *Global Road Safety Factfile 2006.* Retrieved from www.fiafoundation.com/commisionforglobalroadsafety/factfile/index.html.
2. Society of Plastics Engineers (SPE) – Automotive & Composites Division. (2007) Proceeding and presentations from the *7th Annual SPE Automotive Composites Conference* in Troy, Michigan.
3. AFP. (2006). *GM to launch more than 100 Fuel Cell SUV's Worldwide.* World Business Council for Sustainable Development – AFP. Retrieved in September 2006 from www.afp.com.
4. Ahmad, S. and Greene, D. (2004). The Effect of Fuel Economy on Automobile Safety: A Reexamination (TRB-05-1336). Posted at www.ornl.gov.
5. Aksys. (2005). *GMT Composite Pedestrian Beam Wins in New Safety Category at SPE Innovation Award Gala Pedestrian Beam Protects at Lower Weight, Cost.* Retrieved in December 2005 from www.quadrant.ch.
6. Albright, B. (2006, September). *Toyota Debuts Collision Avoidance Systems in Lexus.* (Origin of article is unknown).
7. American Plastics Council. (2001). *Plastics in Automotive Markets Vision and Technology Roadmap.*
8. Lovins, A.B., Datta, E.K., et al. (2004). *Winning the Oil Endgame American Innovation for Profits, Jobs, and Security.* RMI.
9. Argonne National Laboratory. (2005). *DOE Officials Visit Argonne's Transportation Facilities.* Retrieved from www.transportation.anl.gov/features/20050506_bodman_visit.htm.
10. Argonne National Laboratory.(2005). *Recycling Automotive Scrap.*
11. Houser, A. (2005) *Fact Sheet on Older Drivers and Automobile Safety.* AARP Public Policy Institute.
12. ASC. (2006, January). *ASC (American Specialty Cars) Shows off Capabilities in Design, Open Air, Performance Materials Like.* Retrieved from www.ascglobal.com.
13. Associated Press. (2006, September). *GM Adding Hydrogen Fuel Cell Vehicle to California Fleet.*
14. Aucken, A. (2006, March). *Fully Composite Car Leads the Way.* Retrieved from www.advanced-composites.com.
15. Auto Alliance (Alliance of Automobile Manufacturers). *Driving Innovation-* Factsheets posted at www.autoalliance.org:
    a. *Auto Safety Enhancements.*
    b. *Automakers Compatibility Commitment.*
    c. *Improving Everyone's Safety Through Voluntary Industry Cooperation.*
    d. *Alliance Response to NHTSA Electronic Stability Control Announcement.*
    e. *Obsessed with Safety: Your car- a Cocoon of Safety.*
16. Automotive Composites Alliance. (2006). *Latest Developments in Thermoset Composites from the 2006 Automotive Composites Conference & Exhibition.* CD.
17. Automotive Composites Consortium. *Developing Structural Composites for Large Automotive Parts.*

REFERENCES

18. Automotive Design and Production. *Cost-Effective Composite Cars?* Retrieved from www.autofieldguide.com/articles/wip/0505wip02.html.
19. Automotive Learning Center. (2004). *Frequently Asked Questions on Safety*, Retrieved from www.plastics-car.com/s_plasticscar/doc.asp?CID=407&DID=1610.
20. *Automotive Plastics Web site* www.americanplasticscouncil.org/s_apc/sec.asp?CID=303&DID=902. American Plastics Council, Accessed July 2006.
21. Automotive.com. (2006). *2006 Toyota Prius Review: Model Lineup – The environment's best friend.* Retrieved from www.automotive.com/2006/43/toyota/prius/reviews/lineup/index.html.
22. Autoworld. *Composites: Multiple Strengths, Infinite Possibilities.* Sponsored by the Automotive Composites Alliance of the ACMA.
23. Barret, R. (2006, September). Racing Speeds up Auto Innovation. *Boston Sunday Globe.*
24. Braver, E.R., and Trempel, R.E. (2004). Are Older Drivers Actually at Higher Risk of Involvement in Collisions Resulting in Deaths or Nonfatal Injuries Among their Passengers and other Road Users? *Injury Prevention 10*, 27-32.
25. Brosius, D. (2006, June) Engineering Insights: LFRT Uncovers Hidden Value for New Jeep Vehicle. *Composites Technology.*
26. Brown, K. (2006, October). Gentlemen, start your hydrogen rotary, turbo parallel hybrids, what's next; concept cars go green, even using solar panels to turn on the cool air. *FutureCars.*
27. Bunkley, N. (2006, October). Compact Cars May Not be Brawny, but They Are Getting Safer. *The New York Times.*
28. Burkhardt, J, and Eberhard, J. (2003). *Technical Background for a Symposium on Transportation Mobility for the Elderly* (NCHRP Project 20-24[24]).
29. Burkhardt, J. (2003). *Improving Transportation Services – Visions for Mobility.* CalACT Spring Conference, April 2003.
30. Burr, S., Hill, S., and Ramoo, R. (2003). *Determining Accuracy of and Establishing Guidelines for Use of High Strain Rate Polymeric Material Data in Predictive Crashworthiness Models.* Joint Project between SAE, University of Dayton, and Altair Engineering.
31. Busel, J. (2006). *CM Multiple Strengths, Infinite Possibilities: On My Honor: Scouts Need Industry Pros to Support Composite Material Merit Badge.* Composites Manufacturing Association.
32. Buters, J.T., Schober, W. et al. (2007). Toxicity of Parked Motor Vehicle Indoor Air. *Environmental Science and Technology, Vol. 41. No 7*, 2622-2629.
33. Caird, J. (2004). In-vehicle Intelligent Transportation Systems: Safety and Mobility of Older Drivers. *Conference Proceedings 27, Transportation Research Board,* 236-255.
34. Cameron, K. (2006, October). Aluminum is Striving for Mass Appeal, but With a Lot Less Mass. *The New York Times.*
35. Carbon-Fiber. Retrieved in May 2006 from www.answers.com/topic/carbon-fiber.
36. Carpenter, J., Daniels, E., Sklad, P., Warren, D., and Smith, M. (2006). *The R&D of the FreedomCAR Materials Program.* Washington, DC: U.S. Department of Energy.
37. Carpenter, J. (2006). *PNGV/FreedomCAR Materials Crashworthiness R&D.* Washington, DC: U.S Department of Energy.

REFERENCES

38. Chappell, L., and Nussel, P. (2006, August). Weight Watchers: Suppliers to make vehicles lighter. *Automotive News.*
39. Chea, T. (2006). Silicon Valley Entrepreneurs race for Electric Car Market. *San Francisco Chronicle.*
40. Chea, T. (2006, October). Race is on to bring electric sports cars to market. *Boston Sunday Globe.*
41. Cirincione, R. (2006). *Innovation and Stagnation – In Automotive Safety and Fuel Efficiency.* Center for the Study of Responsive Law.
42. Claybrook, J., (2004). President, Public Citizen. *Comments on Side Impact Protection, NPRM 69FR27900 et seq., May 17, 2004.*
43. The Budd Company, Design Center. (2001). *Composite Materials Promise Increased Fuel Efficiency.*
44. Composites: Low-Cost Carbon Fiber – Will the Market Develop? (2006, June). *Composites Technology.*
45. Reinforced Thermoplastics – Thermoformable Composite Panels, Part II. (2006, June). *Composites Technology.*
46. Coughlin, J.F. *Testimony before United States Senate Special Committee on Aging,* April 27, 2004.
47. Coughlin, J., and Reimer, B. (2006). *New Demands form an Older Population: An Integrated Approach to Defining the Future of Older Driver Safety.* SAE Convergence Paper #06SS-51, Massachusetts Institute of Technology.
48. Coughlin, J. (2005). *Disruptive Demographics: Old Age & New Transportation Demands.* New England University Transportation Center.
49. Crash Injury Research and Engineering Network (CIREN) postings at www-nrd.nhtsa.dot.gov/departments/nrd-50/ciren/CIREN.html.
50. Crashworthiness Resource. (2006, July). 2006 Most Recalled Vehicles. *Automobile Manufacturer Recalls and News.*
51. Crashworthiness Resource. (2005, June). GM Issues Saturn Recall. *Automobile Manufacturer Recalls and News.*
52. Crashworthiness Resource. (2006, June). New Technology Could Reduce Rollover Risk. *Automobile Manufacturer Recalls and News.*
53. Crashworthiness Resource. Airbag Failure. *Automobile Manufacturer Recalls and News.*
54. Crashworthiness Resource. (2005, December). Attorney Generals Challenge Roof Strengthening Bill. *Automobile Manufacturer Recalls and News.*
55. Crashworthiness Resource. Auto Accident Facts. *Automobile Manufacturer Recalls and News.*
56. Crashworthiness Resource. Automotive Defects-Roof Crush. *Automobile Manufacturer Recalls and News.*
57. Crashworthiness Resource. (2005, April). GAO Report Confirms Government Test Crash Program Out-Dated. *Automobile Manufacturer Recalls and News.*
58. Crashworthiness Resource. Roof Crush Injury. *Automobile Manufacturer Recalls and News.*
59. Cummings-Saxton, J. (2001). *Automobile Materials Competition: Energy Implications of Fiber-Reinforced Plastics,* Argonne National Lab, ANL/CNSV-25.
60. D'Amico, R. (2006). *Breaking Molds – Bayer Material Science and Rinspeed Develop New Concept Car.* American Composites Manufacturers Association.

REFERENCES

61. Dallmeyer, K. *Fuel Cell Transit Bus, Alternative Fuels, Infrastructure Technologies, Bus Technologies and Testing Information.* Washington, DC: USDOT Federal Transit Administration (FTA).
62. Das, S. et al. (2001). *Evaluation of the Benefits Attributable to Automotive Lightweight Materials Program Research and Development Projects.* Oak Ridge National Laboratory, ORNL/TM-2001-237.
63. Das, S. (2001). *The Cost of Automotive Polymer Composites: A Review and Assessment of DOE's Lightweight Materials Composites Research,* Oak Ridge National Laboratory.
64. DeCicco, J.M. (2005). Environmental Defense. *Steel and Iron Technologies for Automotive Lightweighting.*
65. Decina, L.E. et al. (2003). *Model Driver Screening and Evaluation Program Final Technical Report Volume I: Project Summary and Model Program Recommendation* (DOT HS 809 582). Office of Research and Traffic Records.
66. *Declining death rates due to safer vehicles, not better drivers or improved roadways.* IIHS, August 2006.
67. *Designing Roadways to Safely Accommodate the Increasingly Mobile Older Drives: A Plan to Allow Older Drivers to Maintain Their Independence.* The Roadway Information Program, 2003.
68. Detroit Auto Show. (2006). *2006 Detroit Auto Show – Dodge Challenger Concept.* Retrieved from www.edmunds.com.
69. Detroit Auto Show. (2006). *2006 Detroit Auto Show – Ford Reflex.* Retrieved from www.edmunds.com.
70. Detroit Auto Show. (2006). *2006 Detroit Auto Show – Mitsubishi Concept-CT.* Retrieved from www.edmunds.com.
71. Dittow, C. (2006). *Stabholder – Advocacy Group.* Center for Auto Safety. Retrieved from www.autosafety.org.
72. El Nasser, H. (2006). A Nation of 300 Million. *USA Today.*
73. *Electronic stability control could prevent nearly one-third of all fatal crashes and reduce rollover risk by as much as 80%; effect is found on single – and – multiple – vehicle crashes.* IIHS, June 2006.
74. *Enhancing Future Automotive Safety with Plastics – Technology Integration Report.* Energetics Incorporated. Prepared for the American Plastics Council, May 2006.
75. Evans, L., President, Science Serving Society. *How to Make a Car Lighter and Safer.* SAE TP 2004-01-1172. Retrieved from www.scienceservingsociety.com.
76. Johnson, N.F., Warren, D., Carpenter, J., and Sklad, P. (2004). *FY 2004 Progress Report on Automotive Lightweighting Materials: Composite-Intensive Body Structure Development for Focal Project 3,* Oak Ridge National Laboratory.
77. Fambro, S. (2006). *A 330 MPG Car for Everyone? Three San Diego Engineers Form Company to Build, Sell Revolutionary Car.* Retrieved in January 2006 from www.acceleratedcomposites.com.
78. Farina, A. (2006). *DuPont Launches New Material with Kevlar for Rubber-Based Applications.* Retrieved in July 2006 from www.dupont.com.
79. *Fatality Facts 2004.* (2004). Insurance Institute for Highway Safety.
80. Feraboli, P. (2006). Toward The Development of a Test Standard for Characterizing The Energy Absorption of Composite Materials. *Proceeding of the 6th SPE Automotive Composites Conference,* Warren, Michigan..

REFERENCES

81. Feraboli, P. (2006). Current efforts in standardization of composite materials testing for crashworthiness and energy absorption (AIAA 2006-2217). University of Washington.
82. Feraboli, P. (2006). Current efforts in standardization of composite materials testing for crashworthiness and energy absorption. *Proceeding of the 47th American Institute of Aeronautics and Astronautics Structures, Dynamics and Materials Conference, Newport, Rhode Island*, 2006-2217.
83. Feraboli, P. (2006). Discussion on Issues of Relevance to the NHTSA Safety R&T Roadmap for Future Plastics and Composite Intensive Vehicles (PCIV). University of Washington, private communication.
84. Feraboli, P. (2005). *Minutes of the 1st Crashworthiness Working Group Meeting, CMH-17 Coordination meeting – Charlotte, NC.*
85. Feraboli, P. (2005). *Minutes of the 2nd Crashworthiness Working Group Meeting, CMH-17 Coordination Meeting – Santa Monica, CA.*
86. Feraboli, P. (2006). *Minutes of the 3rd CMH-17 Crashworthiness Working Group Meeting, CMH-17 Coordination Meeting – Chicago.*
87. *FHWA's Enhanced Night Visibility Report Series.* Transportation Research Board, September 2006.
88. Fiat, C.R., (2004). *The Research Requirements of the Transport Sectors to Facilitate an Increased Usage of Composite Materials*, Part II: the Composite Material Research Requirements of the Automotive Industry. .
89. Francfort, J. (2006). *FreedomCAR & Vehicle Technologies Program.* Idaho National Laboratory.
90. Frank, G.J, and Brockman, R.A. (2001).: A Viscoelastic-viscoplastic constitutive model for glassy polymers. *International Journal of Solids and Structures, 38*, 5149-5164. University of Dayton.
91. Frucci, A. (2006). *Honda: Fuel-cell Vehicles on the Road by…(Page cutoff).* Retrieved in June 2006 from www.fuelcellmagazine.com.
92. Gangloff, C. (2006). *Automotive Coatings, Adhesives & Sealants Demand to Exceed $7 Billion in 2010.* Retrieved in April 2006 from www.freedoniagroup.com.
93. Gangloff, C. (2006). *Polyurethane Resin Demand to Reach 7.6 Billion Pounds in 2009.* Retrieved in April 2006 from www.freedoniagroup.com.
94. Gehm, R. (2006, July). Materials Innovations: OEMs Stick with SMC. *Automotive Engineering Journal of Society of Automotive Engineers (SAE).*
95. Gehm, R. Plastic on the Outside. (2006, August). *Automotive Engineering Journal of SAE* 1-114-8-46.
96. General Motors. (2006). *General Motors Recalls 30,000 Corvettes.* Retrieved in May 2006 from www.gm.com.
97. General Motors. *Saturn Vue Sport Utility Specification Sheet.*
98. Gerald, D. (2000). *Extrication Safety: Hybrid Vehicles, Airbags and Moveable Pedals.* Institute for PreHospital Medicine.
99. Gordon, D., Greene, D., Ross, M., and Wenzel, T. (2007). *Sipping Fuel and Saving Lives: Increasing Fuel Economy Without Sacrificing Safety.* Report by October 2006 experts workshop "Simultaneously Improving Vehicle Safety and Fuel Economy Through Improvements in Vehicle Design and Materials."
100. Gottimukkala, R.(2004). *Review Article on Carbon Fiber Composites,* Oak Ridge National Laboratory.

REFERENCES

101. Gould, L. (2004). *Match Analysis with Materials.* Automotive Design and Production. Retrieved from www.autofieldguide.com/articles/040409.html.
102. Harris Interactive. *Advanced Automotive Technologies Assessment – U.S Consumer Acceptance of Advanced Automotive Technologies.* AutotechCAST.
103. Hazen, J. (2006). *Automotive Composites - A Design and Manufacturing Guide – 2nd Edition.*
104. Heudorfer, B., Breuninger, M., Karlbauer, U., Kraft, M., and Maidel, J. *Roofbag-A Concept Study to Provide Enhanced Protection for Head and Neck in Case of Rollover.* Takata-Petri Germany.
105. Hill, H.D., and Ng Boyle, L. (2006). Assessing the Relative Risk of Severe Injury in Automotive Crashes for Older Female Occupants. *Accident Analysis and Prevention, Vol. 38, No. 1,* 148-154.
106. Hill, S., and Sjoblom, P. (1998). Practical Considerations in Determining High Strain Rate Material Properties (University of Dayton Research Institute for SAE Technical Paper Series #981136).
107. Hollowell, W.T. (2006). Overview of NHTSA Research for Enhancing Safety. *USDOT/NHTSA Presentation to the MADYMO International Users' Meeting, Detroit, Michigan.* .
108. *Honda shows off cleaner diesel, streamlined fuel-cell cars.* Retrieved in September 2006 from www.boston.com.
109. Hopkins, H. (2006). *New "Stars on Cars" Rule Announced Today Will Help Consumers Evaluate Safety of New Cars.* Washington, DC: NHTSA.
110. www-nrd.nhtsa.dot.gov/departments/nrd-50/ciren/CIREN.html.
111. Hydrogen, Fuel Cells and Infrastructure Technologies Program www1.eere.energy.gov/hydrogenandfuelcells.
112. Icon aims for a muscular return. *Boston Globe Associated Press.*
113. Insurance Institute for Highway Safety. (2005). Vehicle Incompatibility in Crashes. *IIHS Status Report Special issue, Vol. 40, No. 5.*
114. Ilcewicz, L. (2006). *Composite Safety & Certification Initiatives (Presentation at Fort Worth DER Seminar).* Federal Aviation Administration (FAA).
115. *Improving the Safety of Older Drivers.* (2002). Transportation Association of Canada.
116. *Industry Statistics.* Retrieved in July 2006 from www.americanplasticscouncil.org/s_apc/sec.asp?TRACKID=&SID=6&VID=86&CID=296&DID=895. American Plastics Council.
117. Information Resources www1.eere.energy.gov/vehiclesandfuels/resources/index.html.
118. Inside Line. (2006). *Coming Soon: Honda Unveils Driveable Fuel-Cell Prototype.* Retrieved in September 2006 from www.edmunds.com/insideline/do/News/articleid=116937.
119. Insurance Institute for Highway Safety (IIHS). (2003). *Status Report Vol. 38. No. 3..*
120. Insurance Institute for Highway Safety (IIHS). (1999). *Special Issue: Vehicle compatibility in crashes STATUS REPORT.* Vol. 34 #9.
121. Jackson, R.B. and Schlesinger, W.H. (2004). Curbing the US Carbon Deficit. *PNAS vol. 101, No. 45,* 15627-15629. Retrieved November 9, 2004 from www.pnas.org/doi/101073/pnas.0403631101.
122. Jacob, A., et al. (2003). Reinforced Plastics Automotive Supplement, Ashland Specialty Chemical Company.

REFERENCES

123. Jacob, G., Fellars, J., Simunovic, S., and Starbuck, M. (2003). *Crashworthiness of Automotive Composite Material Systems.* University of Tennessee – Materials Science and Engineering Department and Polymer Matrix Composites Group.
124. Jacob, G., Fellars, J., Simunovic, S., and Starbuck, M. (2001). *Energy Absorption in Polymer Composites for Automotive Crashworthiness.* University of Tennessee – Materials Science and Engineering Department and Polymer Matrix Composites Group.
125. Jeryan, R., Warren, D., and Carpenter, J. (2005). *Automotive Lightweighting Materials – Composite Crash Energy Management.* Ford Research and Innovation Center.
126. Johnson, K., and Mascarin, A. (2002). New Materials Technologies in the Automotive Industry: A Review of Successes and Failures. *SAE International Technical Papers 2002-01-2038.*
127. Kahane, C.J. (2003). *Vehicle Weight, Fatality Risk And Crash Compatibility Of Model Year 1991-99 Passenger Cars And Light Trucks.* Washington, DC: National Highway Traffic Safety Administration. Retrieved from www.nhtsa.dot.gov/cars/rules/regrev/evaluate/pdf/809662.pdf.
128. Kahane, C. (2004). *Lives Saved By The Federal Motor Vehicle Safety Standards And Other Vehicle Safety Technologies, 1960-2002 - Passenger Cars And Light Trucks - With A Review Of 19 Fmvss And Their Effectiveness In Reducing Fatalities, Injuries And Crashes.* National Center for Statistics and Analysis; National Highway Traffic Safety Administration. Retrieved from www.nhtsa.dot.gov/cars/rules/regrev/evaluate/pdf/809833Part1.pdf.
129. Kahn, A. *MIT Group Presents Research on City Car of the Future.* Retrieved from http://web.mit.edu/newsoffice/2004/smartcars.html. September 2004.
130. Karsner, A. Asst. Secretary, USDOE/EERE. Congressional Testimony on the next generation of vehicle and fuels technology, before House Committee on Energy and Air Quality, May 24, 2006.
131. Khaleel, M., Valimont, J., and Carpenter, J. (2005). *Automotive Lightweighting Materials – Structural Reliability of Lightweight Glazing Alternatives.* Pacific Northwest National Laboratory.
132. Klavora, P., and Heslegrave, R.J., (2002). Senior Drivers: An Overview of Problems and Intervention Strategies. *Journal of Aging and Physical Activity, Volume 10, Issue 3,* 322-335.
133. Kos, E. (2006). *CM In-Frastructure: A future Solution to Infrastructure and Fuel Challenges Faced by the Transportation Industry?* American Composites Manufacturers Association.
134. Kos, E. (2006). *CM In-Frastructure: New Hybrid Bridge Deck Process.* American Composites Manufactureres Association.
135. Kreysler, B. (2006). *CM President's Message: Load and Resistance Factor Design: Why Investments now could Mean Revenue Later.* American Composites Manufacturers Association.
136. Krisher, T. DaimlerChrylser is Betting the U.S. has Smartened Up.
137. Krisher, T. High-end Brands Finish at High End of Car Survey.
138. Kurcz, M., Baser, B., Dittmar, H., Sengbush, J., and Pfister, H. (2005). Replacing Steel with Glass-Mat Thermoplastic Composites in Automotive Spare-Wheel Well Applications. *SAE International Technical Papers 2005-01-1678.*

REFERENCES

139. Lackey, E., Hutchcraft, E., Vaughan, J., and Averill, R. (2006). *ZAPPED: Electromagnetic Radiation and Polymer Composites.* American Composites Manufacturers Association.
140. Left Lane News. *BMW moves to plastic for some coupe body panels.* Retrieved from www.leftlanenews.com/2006/04/28/bmw-moves-to-plastic-for-some-3er-body-panels.
141. Li, G., Braver, E.R., and Chen, L.-H. (2003). Fragility versus Excessive Crash Involvement as Determinants of High Death Rates per vehicle-mile of Travel Among Older Drivers. *Accident Analysis and Prevention, John Hopkins University, 35,* 227-35.
142. Lightweight Materials Program. Retrieved from www.ornl.gov/sci/lightmat/Lightweight.html, Oak Ridge National Laboratory.
143. Lovins, A.B. Rocky Mountain Institute (www.rmi.org).
    a. *Advanced Composites: the Car is at the Crossroads,* by M.M. Brylawski and A. B. Lovins, 2000, posted at www.hypercar.com.
    b. *Comments on NHTSA ANPRM Reforming the Automobile Fuel Economy Standards Program,* April 2004.
    c. *Reinventing the Wheels* Guest Editorial, Environmental Health Perspectives, Vol.11 No 4, A218, April 2005 at www.ehponline.org.
    d. *Hypercars, Hydrogen, and the Automotive Transition* by A.B. Lovins and D.R. Cramer, International Journal of Vehicle Design, vol. 35, Nos. ½, 50-84, 2004.
144. Lund, A., O'Neill, B., Nolan, J., and Chapline, J. (2000). Crash Compatibility Issue in Perspective. *SAE Technical Paper Series 2001-01-1378.* IIHS.
145. Lyman, S. et al. (2002). Older Driver Involvements in Police Reported crashes and Fatal Crashes: Trends and Projections. *Injury Prevention 8:116-120.*
146. Lynn, B.C., (2004). Environmental Defense. *Today's Promises, Tomorrow's Cars? - Lessons for FreedomCAR from the Ghosts of Supercars Past (PNGV).*
147. Markos, S. (2006). *Overview of U.S Transportation Vehicle Fire Safety Requirements.* USDOT John A. Volpe National Transportation Systems Center Research and Innovative Technology Administration.
148. Materials Science and Engineering Laboratory (MSEL). Retrieved from www.msel.nist.gov, www.SciTechResouces.gov.
149. Materials Science and Engineering Laboratory, Polymers Division. Retrieved from http://polymers.msel.nist.gov/index.cfm.
150. Matta, F. (2005). *Don't Cross that Bridge Until We Fix It! – Pultruded FRP Reinforcement for Bridge Repair.* American Composites Manufacturers Association.
151. Mayhew, D.R., et al. (2005). *Collisions Involving Senior Drivers: High-Risk Conditions and Locations.* Insurance Institute for Highway Safety.
152. McGuckin, N., and Liss, S. (2005). Aging Cars, Aging Drivers: Important Findings from the National Household Travel Survey, *ITE Journal, Volume 75, Issue 9,* 30-37.
153. Meyer J. (2004). Personal Vehicle Transportation. *Technology for Adaptive Aging, Board on Behavioral, Cognitive and Sensory Sciences and Education,* 253-282.
154. Mills, K., Director, Clean Car Campaign and Haxthausen, E. (2004). Environmental Defense Comments on NHTSA ANPRM Reforming the Automobile Fuel Economy Standards Program.
155. Mintz, J.J. et al. (2000). From Here to Efficiency: Time Lags Between Introduction of Technology and Achievement of Fuel Savings. *Transportation Research Record 1738,* 100-105. TRB, posted by TRISonline.

REFERENCES

156. *MIT Age Lab*, Massachusetts Institute of Technology, http://web.mit.edu/agelab/.
157. Modern Plastics Worldwide. (2006). *Economy & Markets – North American plastics: Fall strength continues.*
158. Modern Plastics Worldwide. (2006). *Modern Executives - Old answers to new questions: dodging the obstacles of plastics innovation.*
159. Moore, R. *Saturn Vehicle Crashes Part 1, 2, and 3.* Retrieved from www.firehouse.com.
160. MSN. (2006). *Toyota sees green in "bioplastics" for cars.* Retrieved from www.msnbc.msn.com/id/4637563/.
161. Nagourney, E. (2005, October). Ideas for Making it Easier to Walk Away From an Accident. *New York Times,* p. 34.
162. Naravane, A., Deb, A., Shivakumar, N., and Chiottappa, H. (2005). *A Comparative Study of Composite and Steel Front Rails for Vehicle Front Impact Safety.* The Automotive Research Association of India 2005-26-324.
163. National Research Council. (2005). *Review of the Research Program of the FreedomCAR and Fuel Partnership: First Report.*
164. National Transportation Research Center. FY 2002-2005 ORNL Transportation Program Highlights. Retrieved from www.ntrc.gov/transportation2005.shtml.
165. New Plastics and the Automobile. *SAE International Journals 1-108-5-70,* May 2000.
166. NHTSA. *Traffic Safety Facts 2005.* Retrieved from www-nrd.nhtsa.dot.gov/Pubs/TSF2005.PDF.
167. NHTSA. (1997). *2020 Report: People Saving People- On the Road to a Healthier Future.*
168. NHTSA Studies and Reports (on Web site): http://nhtsa.gov/portal/site/nhtsa/menuitem.a8131659c3c0a2381601031046108a0c/;jsessionid=EMyPS4zjvEasDAYBAlHs1Qm9ncn71vGM2plOk6RHZGrZD1YKNjuK!28012360.
169. *NHTSA Vehicle Safety Rulemaking and Supporting Research Priorities for CY 2005-2009* is posted at www.nhtsa.gov/cars/rules/rulings/PriorityPlan-2005.html and the summary for Congress is at www.nhtsa.dot.gov/nhtsa/announce/NHTSAReports/PriorityPlan-2005.html.
170. NHTSA. (2005). *Consequences and Costs of Lower Extremity Injuries* (DOT HS 809 871). Office of Vehicle Safety Research, National Highway Traffic Safety Administration.
171. NHTSA. (2004). *NHTSA's Four Year Plan for Hydrogen, Fuel Cell and Alternative Fuel Vehicle Safety Research.* National Highway Traffic and Safety Administration.
172. Noran Engineering. *NEiLaminate Tools – Nastran Finite Element Analysis Software.*
173. Notices, Federal Register, Vol. 71, No. 133, July 2006.
174. Oak Ridge National Laboratory Transportation Program. (2005). *Carbon Fiber Systems Integration.* Retrieved from www.ntrc.gov/pdfs/transportation2005/14_ALM_1.pdf.
175. Oak Ridge National Laboratory. (2005). *Aid for the Auto Industry.* Retrieved from www.ornl.gov/info/ornlreview/v38_1_05/article05.shtml.
176. Oak Ridge National Laboratory. (2003). *Automotive Lightweighting Materials, Low-cost Carbon Fiber from Renewable Resources.*
177. Oak Ridge National Laboratory. (2005). *Multiple Roads to the Hydrogen Car,* www.ornl.gov/info/ornlreview/v38_1_05/article06.shtml.

REFERENCES

178. Oak Ridge National Laboratory. (2004). *Composites Durability.* Retrieved from www.ntrc.gov/pdfs/transportation2004/Composites_Durability.pdf.
179. Oak Ridge. (2006). *Carbon Fiber Cars Could Put U.S. on Highway to Efficiency.* Retrieved in March 2006 from www.ornl.gov/news.
180. Ohio State University. *Center for Advanced Polymer and Composite Engineering* funded by the National Science Foundation.
181. Older Drivers Up Close Aren't Dangerous Except Maybe to Themselves. (2001). *Status Report, Insurance Institute for Highway Safety, Volume 36, Number 8.*
182. Olson, R. (2004). *Senior Driver Issues: Upcoming Challenges and Solutions.* International Risk Management Institute, Inc.
183. Balaguru, P.N., et al. (1998). Fire-resistant aluminosilicate composites. *Fire and Material, Volume 2, Issue 2,* 67- 73.
184. Pfestorf, M., and Rensburg, J. (2006). Improving the Functional Properties of the Body – In-White with Lightweight Solutions Applying Multiphase Steels, Aluminum and Composites. *SAE International Technical Papers 2006-01-1405.*
185. Pickrell, D. (2002). *Characteristics of the Future Elderly Population and their Implications for Travel Behavior.* Volpe National Transportation Center White Paper.
186. Pinell, M. (2005). *Rough Order of Magnitude (ROM) Cost for the High Strain Rate Test Sample Study for Long Fiber Filled Polymers.* E-mail from M. Pinell to Gary Pollack, June 2005.
187. Pinnell, M. (2006). *Special Concerns in High Strain Rate Tensile Testing of Polymers.* University of Dayton for SAE International 2006-01-0121.
188. Pittle, D., Senior VP, Technical Policy, Consumers Union. *Testimony, on Reauthorization of the National Highway Traffic Safety Administration before the Subcommittee on Commerce, Trade and Consumer Protection of the House Committee on Energy and Commerce,* March 2004.
189. Pittle, D. (2004). *Comments of Consumers Union of the U.S. Inc. to the Department of Transportation National Highway Traffic Safety Administration – In Response to Advance Notice of Proposed Rulemaking 49CFR Part 533 Docket No. 2003-16128, 2127-AJ17 on Reforming the Automobile Fuel Economy Standards Program.*
190. Plastics: The Return of Body Panels? (2006). *Automotive Design and Production.*
191. Pope, B. (2006). *WARDS'S AutoWorld: Ford Belt Driven.* Primedia Inc.
192. *Program Management for Automotive Lightweighting Materials.* Retrieved from www.ms.ornl.gov/programs/energyeff/lwvm/default.htm. U.S. Department of Energy Office of FreedomCAR and Vehicle Technologies.
193. Red, C. (2006). *2025 Advanced Composites – Market Trends for the Next 20 Years.* American Composites Manufacturers Association.
194. Reinforced Plastics. (2004). *Carbon Car Panels a Cost Effective Reality.* Retrieved from www.reinforcedplastics.com/market_focus/automotive/spfeature.html.
195. Reinforced Plastics. (2004). *Composite Bonnet Suits Low Volume Sports Car.* Retrieved from www.reinforcedplastics.com/market_focus/automotive/azdel.html.
196. Reinforced Plastics. (2004). *Hybrid front-end for For Focus C-MAX van.* Retrieved from www.reinforcedplasticsbuyersguide.com/WZ/RPlastics/news/applications_news/000088/show.
197. Ressler, A. (2004). *Crash the Crash Test Dummy, An Analysis of Individual Factors in Fatal Car Crashes.* University of Pennsylvania – ESE202.

REFERENCES

198. Richard, T. (2006). Aging Drivers: Storm in a Teacup? *Accident Analysis and Prevention, Volume 38, Issue 1.*
199. Robinson, E. (2006). Resin Prices Expected to Remain High. *Automotive News.*
200. Rocky Mountain Institute. *Advanced Composites, Advanced Composites Have Many Benefits, Ultralight Construction, HyperCar Design and Technology, The HyperCar Concept, Safety, Fuel Savings, How to Design a Better Road Vehicle.* Retrieved from www.rmi.org.
201. Rocky Mountain Institute. *Transportation: Hypercar Chronology – Advanced Automotive Milestones - Elements of Hypercar Vehicles are Emerging.* Retrieved from www.rmi.org/sitepages/pid389.php.
202. Ross, M., and Patel, D. (2006). *Intrusion in Side Impact Crashes* for SAE International. University of Michigan Physics Department.
203. Ross, M., and Wenzel, T. (2001). *Losing weight to save lives: a review of the role of automotive weight and size in traffic fatalities.* American Council for Energy-Efficient Economy, for DOE/LBL (ACEEE-T013). University of Michigan Physics Department and Lawrence Berkeley Laboratory.
204. Ross, M., Patel, D., and Wenzel, T. (2006, January). Vehicle Design and the Physics of Traffic Safety. *Physics Today.*
205. Ross, M., Patel, D, Compton, C., and Wenzel, T. (2006). *Intrusion and the Aggressivity of Light Duty Trucks in Side Impact Crashes.* University of Michigan Physics Department, University of Michigan Transportation Research Institute, and Lawrence Berkeley National Laboratory.
206. Rumy, Z. *Zoltek Partners with BMW.* Retreived from www.about.com.
207. Making Composites for HANS at Nascar is Part Work, Part Hobby. (2006, March). *Salt Lake Tribune.* Retrieved from www.sltrib.com/jordan/ci_3558872.
208. Saturn Vehicle Crashes, Part 1. Retrieved from www.Firehouse.com.
209. Sawyer, C.A. (2005). *Plastic – Intensive Vehicle Déjà vu?* Retrieved from www.autofieldguide.com/articles/060506.html. Automotive Design and Production.
210. Sayers, M. (2006). *Advances in LED Automotive Forward-Lighting: Providing Catalysts for Change.* Viseteon Corp.
211. Schroder, S. (2004, February). Porsche Engineering. *Porsche Engineering Magazine.*
212. Schweitzer, J. (2006). *Managing the Business Risk of Styrene.* American Composites Manufacturers Association.
213. Seats, R.L, Fisher, D., and Twardowska, H. *Tough, Low Mass SMC Development for Transportation Applications.* Ashland, Inc.
214. Seemann, B. (2006). *Composites Manufacturing.* American Composites Manufacturers Association.
215. Shane, J. (2007). USDOT Undersecretary of Transportation Policy, Congressional Testimony on The Surface Transportation System: Challenges for the Future, January 24, 2007.
216. Sherman, L.M. (2004, December). Polyurethanes. *Plastics Technology.*
217. Shopping for Safety: Providing Customer Automotive Safety Information. (1996). *Special Report No. 248.* Transportation Research Board.
218. Siemens AG Germany. (2004, November). Hybrid Material has Metallic Properties, Processed like Plastic. *Advanced Composites Bulletin.*

REFERENCES

219. Simunovic, S., Starbuck, M., and Carpenter, J. (2005). *Automotive Lightweighting Materials - Modeling of High Strain-Rate Deformation of Steel Structures.* Oak Ridge National Laboratory.
220. Skinner, D., and Stearns, M.D. (1999). *Safe Mobility in an Aging World,* John A. Volpe National Transportation Systems Center, U.S. Department of Transportation.
221. Society of Plastics Engineers. (2006). *Proceedings of the 6th Annual SPE Automotive Composites Conference – Composites: Efficiency, Value, Performance.* September 2006.
222. Society of Plastics Engineers. *Automotive Plastics: Adding Functionality, Reducing Vehicle Cost.*
223. Starbuck, M., Warren, D., and Carpenter, J. (2005). *Automotive Lightweighting Materials – Intermediate-Rate Crush Response of Crash Energy Management Structures.* Oak Ridge National Laboratory.
224. Starbuck, M., Rastogi, N., Warren, D. and Carpenter, J. (2005). *Automotive Lightweighting Materials – Crash Analysis of Adhesively-Bonded Structures (CAABS).* Oak Ridge National Laboratory.
225. Starnes, K. (2006). *Jeep Compass Achieves Durability, Stiffness While Reducing Noise, Vibration, Harshness.* Retrieved in June 2006 from www.dcx.com.
226. Stone, J.L., President, Advocates for Highway & Auto Safety. 2002 statements.
227. Stutts, J. (2005). *NCHRP Synthesis of Highway Practice Issue 348*: Improving the Safety of Older Road Users, Transportation Research Board.
228. Sullivan, R. (2006). *FreedomCAR and Vehicle Tecnology Program.* Advanced Materials and Vehicle Systems, *USDOE,* -presentation at ACCE-06, SPE Automotive Composites conference.
229. Swart, C. (2006). *DOW Automotive Wins 2006 PACE Award.* Retrieved in April 2006 from www.dowautomotive.com.
230. Teaching tools. (2003). *How are Plastics Made?* Retrieved from www.teachingtools.com/Slinky/plastics.html.
231. Technologies www1.eere.energy.gov/vehiclesandfuels/technologies/materials/index.html.
232. Tessier, C. (2006). *GE Plastics Intensifies Global Application Technology (GApT) to Leap-Frog Creation of Next-Gen Apps.* Retrieved in April 2006 from www.geplastics.com.
233. *The CFRP Automobile Project in Japan* by A. Kitano, E. Wadahara and I. Taketa, Paper presented at at the 12th US-Japan Conference on Composite Materials, September 21-22, University of Michigan-Dearborn, MI.
234. The Ecology Center, HealthyCar.org- reports on toxic Volatile Organic Chemicals (VOC) and other chemicals in vehicles:.
    a. *Toxic at Any Speed-Chemicals in cars and the need for Safe Alternatives.*
    b. *2006 Automotive Plastics Report Card-The Policies and Practices of 8 Leading Automakers.*
    c. *Flame retardants: Alarming Increases in Humans and the Environment*.
235. *The Inside Story: Why Interiors Matter.* (2006). Automotive Design and Production. www.autofieldguide.com/articles/060603.html.
236. Tomita and Tsutomu. (2006). *Toyota Launches TF106 Formula One Car.* Retrieved in January 2006 from www.toyota-fl.com/public/en/technologies.tf106.html.

REFERENCES

237. *Tomorrow's Materials: Lighter, Tougher, Faster.* Retrieved from www.nsf.gov/about/history/nsf0050/materials/tomorrow.htm, National Science Foundation.
238. *Transportation in an Aging Society. Improving Mobility and Safety for Older Persons, Volume 1 – Committee Report and Recommendations,* TRB, 1998.
239. *Transportation in an Aging Society. Improving Mobility and Safety for Older Persons, Volume 2 – Technical Papers,* Report HS-040 567, TRB, 1998.
240. *Transportation in an Aging Society-A Decade of Experience,* Transportation Research Board Conference Proceedings 27, TRB, 2004.
241. Transportation Research Board. *Improving Public Transit Options for Older Persons – Volume 2 Final Report.* TRB's Transit Cooperative Research Report 82, February 2003.
242. U.S. Department of Energy (DOE). (2003). *FY 2003 Progress Report for High Strength Weight Reduction Materials,* Energy Efficiency and Renewable Energy.
243. U.S. Department of Energy (DOE). (2001). Office of Transportation Technologies. *Composite Materials Production Methods Developed, Transportation for the 21st Century.*
244. U.S. Department of Transportation. (2006). *Strategic Research, Development and Technology Plan,* 2006-2010, RITA-2006-25247-1.
245. U.S. DOT. (2006). *Department of Transportation Draft Strategic Plan Fiscal Years 2006-2011.*
246. U.S. DOT. (2006). *Department of Transportation Strategic Plan Fiscal Years 2006-2011 – Safety Strategic Goal.*
247. United States Army. (2006). *Small Business Innovation Research (SBIR) Program.* Department of Defense Automotive Resources.
248. United States Government Accountability Office (GAO). (2005). *Vehicle Safety – Opportunities Exist to Enhance NHTSA's New Car Assessment Program.* GAO.
249. United States Government Accountability Office. (2004). *Transportation – Disadvantaged Seniors: Efforts to Enhance Senior Mobility Could Benefit from Additional Guidance and Information.* GAO.
250. University of New Mexico, Rutgers, TCNJ. *Ceramic and Composite Materials Center (CCMC).* National Science Foundation (NSF).
251. USDOC Census Bureau. *65+in the United States: 2005.* Retrieved from www.census.gov.
252. USDOE *2006 Strategic Plan.* Retrieved from www.energy.gov.
253. USDOE and USDOT. (2006). *Hydrogen Posture Plan-An Integrated Research, Development, and Demonstration Plan.* Retrieved from www1.eere.energy.gov/hydrogenandfuelcells.
254. USDOE Public Affairs. (2005). *DOE Provides $4.7 Million to Support Excellence in Automotive Technology Education.*
255. USDOE, FreedomCAR and Fuel Partnership Materials Technology Roadmap 2006 at www1.eere.energy.gov/vehiclesandfuels/pdfs/program/materials_team_technical_roadmap.pdf.
256. USDOE. (2006). *Driving Technology: A Transition Strategy to Enhance Energy-A Transition Strategy to Enhance Energy Security.* 2006-08-01 Retrieved from www.eere.energy.gov/vehiclesandfuels/pdfs/program/tsp_paper_final.pdf.

REFERENCES

257. USDOE. (2006). *FreedomCAR and Fuel Partnership- Materials Technology Roadmap.* Retrieved from www1.eere.energy.gov/vehiclesandfuels.
258. USDOE. (2006). *FreedomCAR and Fuel Partnership- Vehicle Systems Analysis Technical Team.* Retrieved from www1.eere.energy.gov/vehiclesandfuels.
259. USDOE. (2005). *FreedomCAR and Fuel Partnership 2005 Highlights of Technical Accomplishments,* FreedomCAR and Fuel Partnership.
260. USDOE. (2005). *Heavy Vehicle Materials Strategy Presentation,* Heavy Vehicle Materials Strategy.
261. USDOE/USCAR Automotive Composites Consortium (ACC). (2005). *Advanced Lightweight Materials (ALM)- Composite Materials- 5 Year Plan.*
262. *USDOE/USCAR FY 2004 Progress Report for High Strength Weight Reduction Materials* – Energy Efficiency and Renewable Energy, 2004.
263. *USDOE/USCAR FY 2004 Progress Report on Automotive Lightweighting Materials: Structural Reliability of Lightweighting Glazing Alternatives,* Oak Ridge National Laboratory, 2004.
264. *USDOE/USCAR FY 2004 Progress Report on Automotive Lightweighting Materials: Composite-Intensive Body Structure Development for Focal Project 3,* Oak Ridge National Laboratory, 2004.
265. USDOT National Highway Traffic Safety Administration. (2006). *Transportation Secretary Mineta Calls Highway Fatalities Tragedy, Says All Americans Can do More to Improve Road Safety.*
266. USDOT Research and Innovative Technology Administration. (RITA). *Transportation Research, Development and Technology Strategic Plan 2006-2010, Section 3: Safety Research, Development, and Technology Priorities.* Retrieved from http://rita.dot.gov/publications/transportation_rd_t_strategic_plan/html .
267. USDOT/FHWA. (2006). *FHWA's Enhanced Night Visibility Report Series.* Transportation Research Board.
268. USDOT/NHTSA. (2005). *NHTSA Vehicle Safety Rulemaking and Supporting Research Priorities.*
269. Vasilash, G.S. (2005). *Creating Clever Things with Plastics.* Retrieved from www.autofieldguide.com/articles/120501.html. Automotive Design and Production.
270. Vasilash, G.S. (2004). *High Velocity Materials Improvement.* Retrieved from www.autofieldguide.com/articles/030407.html, Automotive Design and Production.
271. Vasilash, G.S.. (2004). *Plastics Possible Future.* Retrieved from www.autofieldguide.com/columns/gary/1002mat.html. Automotive Design and Production.
272. Vasilash, G.S. *Saturn Redefined.* Automotive Design and Production.
273. Vasilash, G. *Advanced Composites for an Advanced Corvette.* Automotive Design and Production. Retrieved from www.autofieldguide.com/articles/article_print1.com.
274. Vasilash, G. (2003). *Looks Great. Less Weighty.* www.autofieldguide.com/articles/article_print.cfm.
275. Vasliash, G. (2006). *Plastics: The Return of Body Panels?* Retrieved from www.autofield.com.
276. Villano, P., Meuleman, D., Simunovic, S., and Carpenter, J. (2005). *Automotive Lightweighting Materials – Strain Rate Characterization.* Auto/Steel Partnership (A/SP).

REFERENCES

277. Villano, P., Catterall, J., Shaw, J., and Carpenter, J. (2005). *Automotive Lightweighting Materials – Future Generation Passenger Compartment.* Auto/Steel Partnership (A/SP).
278. Wang, S., and Rupp, J.D. (2006). *Alterations in Body Composition and Injury Patterns with Aging.* University of Michigan Trauma Burn Center and Transportation Research Institute. Presentation posted on the NHTSA Crash Injury Research and Engineering Network (CIREN) Web site at www-nrd.nhtsa.dot.gov/departments/nrd-50/ciren/CIREN.html.
279. Wang, S.C. *An Aging Population: Fragile, Handle With Care.* University of Michigan Trauma Center. Posted on CIREN Web site at www-nrd.nhtsa.dot.gov/departments/nrd-50/ciren/um_fragile.html.
280. Warner, F. (2006, October). From Average-Joe Subcompact to Spymobile. *The New York Times.*
281. Warren, D., Shaffer, J., Paulauskas, F., and Abdullah, M. (2002). Low-Cost Carbon Fiber for the Next Generation of Vehicles; Novel Technologies. *SAE International Technical Papers 2002-01-1906.*
282. Wenzel, T., and Ross, M. (2005). The Effects of Vehicle Model and Driver Behavior on Risk. *Accident Analysis and Prevention, 37,* 479-979).
283. Wheelan, M., Langford, J., Oxley, J., Koppel, S., and Charlton, J. (2006). The Elderly and Mobility: A Review of the Literature. *Monash University Accident Research Centre Report No. 255.*
284. Whitfield, K. (2004). *Do Plastic Body Panels Have a Future?,* Retrieved from www.autofieldguide.com/articles/060407.html. Automotive Design and Production.
285. Wikipedia. (2006). Car Safety. Retrieved from http://en.wikipedia.org/wiki/Car_safety.
286. Wikipedia. (2006). Crash Test Dummy. Retrieved from http://en.wikipedia.org/wiki/Crash_test_dummies.
287. Wikipedia. (2001). *Plastics.* Retrieved from http://en.wikipedia.org/wiki/Plastics.
288. Wikipedia. (2006). Saturn (automobile). Retrieved from http://en.wikipedia.org/wiki/Saturn_%28automobile%29.
289. World Health Organization (WHO). (2004). *World Report on Road Traffic Injury Prevention.* Factsheets on *Road Safety: a Public Health Issue* and *Road Traffic Injury Prevention Training Manual* (2006) posted at www.who.int/violence_injury_prevention/publications/road_traffic/en.
290. Wright, S. (2005). *Future of the Car Talk Highlights MISTI Week.* Retrieved in September 2005 from http://web.mit.edu/newsoffice/2005/misti-cars.html.
291. Yamaguchi, J. (2000). *Automotive Engineering International Online: Shrinking Electric Cars.* SAE International.
292. Zimmons, J. (2006). *Zap Showcases Brazilian Obvio! Alcohol-Gas-Electric "Trybrid" Microcar at New York Auto Show.* Retrieved in April 2006 from www.zapworld.com.

DOT HS 810 863
November 2007

U.S. Department
of Transportation
**National Highway
Traffic Safety
Administration**